Workbook

Manufacturing Execution Systems

An Operations Management Approach

Second Edition

AF871125

Thomas Seubert
Grant Vokey

ISA™

Notice

The information presented in this publication is for the general education of the reader. Because neither the author nor the publisher has any control over the use of the information by the reader, both the author and the publisher disclaim any and all liability of any kind arising out of such use. The reader is expected to exercise sound professional judgment in using any of the information presented in a particular application.

Additionally, neither the author nor the publisher has investigated or considered the effect of any patents on the ability of the reader to use any of the information in a particular application. The reader is responsible for reviewing any possible patents that may affect any particular use of the information presented.

Any references to commercial products in the work are cited as examples only. Neither the author nor the publisher endorses any referenced commercial product. Any trademarks or trade names referenced in this publication, even without specific indication thereof, belong to the respective owner of the mark or name and are protected by law. Neither the author nor the publisher makes any representation regarding the availability of any referenced commercial product at any time. The manufacturer's instructions on the use of any commercial product must be followed at all times, even if in conflict with the information in this publication.

The material and information contained in this book are for general information purposes only. Views and opinions expressed by the author(s) are solely their own and do not necessarily represent those of ISA.

Copyright © 2023 International Society of Automation (ISA)
All rights reserved.

Printed in the United States of America.
Version: 1.0

ISBN-13: 978-1-64331-216-3 (print)
ISBN-13: 978-1-64331-217-0 (ePub)
ISBN-13: 978-1-64331-218-7 (Kindle)

No part of this work may be reproduced, stored in a retrieval system, or transmitted in any form or by any means, electronic, mechanical, photocopying, recording, or otherwise, without the prior written permission of the publisher.

ISA
3252 South Miami Blvd, Suite 102
Durham, NC 27703

Library of Congress Cataloging-in-Publication Data
Names: Seubert, Tom | Vokey, Grant.
Title: Manufacturing execution systems : an operations management approach/Tom Seubert, Grant Vokey.
Description: Second edition | Durham : International Society of Automation, 2023. | Includes bibliographical references and index.
Identifiers: LCCN 2023943101 (print) | ISBN: 9781643312163 (paperback)
LC record available at https://lccn.loc.gov/2023943101

Part I

Exercises

1 Introduction, Overview, and the Current State of Manufacturing

1.1 What is ERP?

What does ERP stand for, and what are the high-level processes?

1.2 What is PLM?

What does PLM stand for, and what are the high-level processes?

1.3 What is MES?

What does MES stand for, and what are the high-level processes?

____

____

____

____

____

1.4 What is MOM?

What is MOM, and what are the four main areas of operations that MOM covers?

____

____

____

____

____

1.5 Manufacturing Enterprise Planning

What are the five main areas of manufacturing enterprise planning within the plant?

1. ________________________________
2. ________________________________
3. ________________________________
4. ________________________________
5. ________________________________

1.6 ISA-95 Level Order

Assign the activities to the appropriate ISA-95 levels.

Level		Activities
Level 0	______	A. Production activity
Level 1	______	B. Control and monitoring
Level 2	______	C. Business planning and logistics
Level 3	______	D. Sensors and devices that support production
Level 4	______	E. Manufacturing operations management

2 A Functional Overview of MES

2.1 Level 3 and Level 4 General Interfaces

Identify the general blocks of data that are included in the interfaces between Level 3 and Level 4.

2.2 Level 2 and Level 3 General Interfaces

Identify the general blocks of data that are included in the interfaces between Level 2 and Level 3.

2.3 Difference between ERP and MES

What is the primary difference between ERP transaction processing and MES transaction processing? Why is this difference significant?

2.4 MES Functionality

List and briefly describe the three primary categories of MES functionality.

3 Order Execution Management: Details and Benefits

3.1 Benefit of MES to Order Execution

Identify a primary value that MES can provide to production planners.

____

____

____

3.2 Other Benefits of MES

Identify other general areas of interest and explain the value that MES provides in the context of order execution.

____

____

____

____

____

4 Material Management: Details and Benefits

4.1 Material/Part Lifecycle

Place the lifecycle stages of the material or part in order.

1.	________	A.	Plant receives the part.
2.	________	B.	Parts are picked in the order needed by production.
3.	________	C.	Parts are consumed and removed from plant inventory.
4.	________	D.	Vendor ships the part.
5.	________	E.	Plant orders the part.
6.	________	F.	Part is transferred to the plant warehouse.
7.	________	G.	Parts are delivered to the production line.
8.	________	H.	Parts are transferred from bulk storage to the shopping area.

4.2 Material State

Place the material states in order, and indicate which states are considered to be part of the in-plant inventory.

1. ________	A. In the warehouse
2. ________	B. Consumed
3. ________	C. WIP
4. ________	D. Ordered
5. ________	E. In the shopping market
6. ________	F. Received

4.3 Material Management Methods

Describe the following material management methods:

- **E-kanban –** ______________________________

  ____
  ____
  ____
  ____

- **Kitting –** ______________________________

  ____
  ____
  ____
  ____

- **Subassemblies –** ______________________________

  ____
  ____
  ____
  ____

- **Bulk parts –** ______________________________

  ____
  ____
  ____
  ____

4.4 Automated MES Material Management

What effect does using an automated MES material management function have on inventory management in general?

____

____

____

____

____

4.5 Selecting the Proper Benefits

Select the benefits that an MES provides for material management.

- ☐ It supports just-in-time production scheduling.
- ☐ It minimizes in-plant inventory, which can be a taxable commodity.
- ☐ It minimizes scrap.
- ☐ Inventory is more accurate.
- ☐ Inventory is leveled to maximize production while minimizing inventory.
- ☐ It directly supports the kitting process by automating the replenishment notification. It also helps minimize selecting incorrect parts and causing end-product quality issues.
- ☐ It helps ensure proper part picking at the production line side.
- ☐ It streamlines the subassembly process to minimize defects before the final assembly process.
- ☐ It automates the ordering functionality and in-plant transfers. Ordering and transfers no longer require a manual process that may not be triggered at the proper time.
- ☐ It annunciates equipment faults.
- ☐ It provides the ability to track parts and products through the manufacturing lifecycle.

- ☐ It supports real-time fault annunciation of all different material management faults that negatively affect production. This speeds up resolution time.
- ☐ It automatically backflushes parts, which in turn triggers payment. Vendor payments no longer require a manual process that may not be triggered at the proper time.
- ☐ It helps define the order schedule.
- ☐ It ensures that parts are sequenced properly when picked from the shopping area.

5 Quality and Traceability: Details and Benefits

5.1 Quality Control versus Quality Assurance

Explain the difference between quality control and quality assurance.

5.2 What is SPC?

What does SPC stand for? How does it support quality assurance in an MES context?

5.3 Poka-yoke

Identify two of the poka-yoke methods used in MES.

1. ______________________________
2. ______________________________

5.4 MES traceability

List four aspects of manufacturing where the traceability provided by MES can be of value to a manufacturing company, and briefly explain why it is of value.

1. ______________________________
____
____
2. ______________________________
____
____
3. ______________________________
____
____
4. ______________________________
____
____

Why is traceability an important aspect of quality?

____
____
____
____
____
____

6 Production Monitoring: Details and Benefits

6.1 Fault Type and Priority

Classify the following faults by type and priority. Fault types include Equipment, Production, and Quality. Priorities include Critical, Major, and Minor. Assume there are no redundant systems used in this example.

Fault	Description	Type	Priority
Workstation A1 main power off	Main power is off at workstation		
Lift station X3 motor overload	Motor is overloaded and not working		
Buffer zone C10 full	Buffer zone near column C10 is full		
Workstation C3 overcycle	Workstation C3 has had an overcycle		
Workstation E50 position sensor needs calibrating	Sensor needs calibration		
Workstation H5 part bin 2 is empty	Part bin containing the subpart for assembly is empty		
Workstation H5 part bin 3 is full	Part bin containing the subpart for assembly is full and cannot accept any more parts		
Line A2B temperature sensor 1 needs calibration	Check calibration		
Line B2C proximity sensor is faulted	Sensor checks for proper part alignment		

Fault	Description	Type	Priority
Workstation XYZ overcycle	Cycle time exceeded		
Workstation V1A misassembly	Assembly process in the station was incorrect		
Inspection station Q2A: part failed stress test	Material failure		
Inspection station FQ: part failed operational test	Part did not operate as per spec		

6.2 Equipment Trend Case Study

Review the following equipment trend study of the top 10 equipment faults. Pick the top three and explain the reasons for your selections.

Top 10 Equipment Faults Report for One Month

Fault	Description	# of Occurrences	Average Downtime	Total Downtime	Replacement Effort
Motor overload	Motor stopped due to overload	10	8	80	Replace motor control relay Time: 45 minutes Cost: $250 per relay
Temp sensor fault	Sensor stopped working	12	5	60	Replace sensor Time: 20 minutes Cost: $500
Pump fault	Pump is not producing flow	2	30	60	Replace pump Time: 30 minutes Cost: $220
Transfer gate stuck open	Part gate stuck open	25	2	50	Replace gate actuator Time: 35 minutes Cost: $120
Transfer gate stuck closed	Part gate does not open	20	2	40	Replace gate actuator, Time: 35 minutes Cost: $120

Fault	Description	# of Occurrences	Average Downtime	Total Downtime	Replacement Effort
Safety barrier breached	When breached, equipment shuts down for safety	20	1.5	30	Replace light screen if necessary Time: 15 minutes Cost: $500
Actuator failed to move	Electrical actuator	1	25	25	Replace actuator Time: 20 minutes Cost: $400
Press fault	Press failed to achieve pressure	2	12	24	Tighten connectors, replace hoses Time: 15 minutes per hose Cost: $50 per hose
Hydraulic pressure fault	Failed to achieve starting pressure	2	11	22	Tighten connectors, replace hoses Time: 15 minutes per hose Cost: $50 per hose
Conveyor B4 drive chain fault	Chain from the drive gear to conveyor broken	1	20	20	Replace chains Time: 20 minutes per chain Cost: $80 per chain

Note: Downtime is in minutes.

Reason table

The list should be prioritized based on the highest downtime impact with the lowest cost and repair time.

Item	Fault	Reason
1		Downtime: _______ Repair time: _______ Cost: _______
2		Downtime: _______ Repair time: _______ Cost: _______
3		Downtime: _______ Repair time: _______ Cost: _______

6.3 Cycle Time for Improvement

Make three suggestions to improve productivity based on the cycle time trend for a manufacturing line that is serial: Station 1 directly feeds Station 2, which feeds Station 3, which feeds Station 4, which feeds Station 5. There are no parallel operations. After an assembly has entered Station 1, it cannot come out until Station 5. The line must produce one part per minute, so the cycle time for each station must be 60 seconds or less.

Cycle	1	2	3	4	5
1	59.5	50.0	61.2	56.0	59.5
2	59.0	45.9	65.0	56.0	59.0
3	58.0	51.5	64.8	56.0	58.0
4	59.9	56.0	64.9	56.1	59.9
5	60.0	47.8	65.2	56.1	60.0
6	58.5	46.7	65.1	56.0	58.5
7	59.0	49.2	66.0	56.1	59.0
8	59.3	53.3	64.4	56.2	59.3
9	59.5	54.4	64.9	56.1	59.5
10	59.6	55.5	64.5	56.0	59.6

1. ____

   ____

2. ____

   ____

3. ____

   ____

6.4 OEE Case Study for Improvement

The following is an OEE report for a line for one shift. Calculate the return on investment (ROI) to raise both the production performance and the first-pass quality rate by 0.05 at an investment cost of $50,000 if each part produced generates revenue of $2000 and costs an average of $1200 per part to repair before shipping.

OEE Calculation per Hour Shift = 8 hours									
Scheduled JPH		60							
Hour	Downtime	Availability	Scheduled JPH	Actual JPH	Performance	Actual Quality	Quality Rate	OEE	Reasons
1	15	0.750	60	40	0.667	35	0.875	0.438	Equipment Down for Repair
2	0	1.000	60	50	0.833	45	0.900	0.750	
3	0	1.000	45	40	0.889	35	0.875	0.778	15 minute break
4	0	1.000	60	51	0.850	45	0.882	0.750	
5	0	1.000	30	25	0.833	22	0.880	0.733	30 minute lunch
6	0	1.000	60	51	0.850	45	0.882	0.750	
7	0	1.000	45	40	0.889	35	0.875	0.778	15 minute break
8	0	1.000	60	53	0.883	46	0.868	0.767	
Shift Total:	15	0.969	420	350	0.833	308	0.880	0.710	

7 Continuous Improvement: Details and Benefits

7.1 Process Improvement Steps

Place the following steps in order.

1. _______	A. Determine the most significant issue for the process.
2. _______	B. Make process improvements to remove (or control) the most significant issue.
3. _______	C. Repeat step 1.
4. _______	D. Characterize the stable process.
5. _______	E. Stabilize the process.
6. _______	F. Determine the root cause of the most significant issue.
7. _______	G. Evaluate the effectiveness of the change.

7.2 CI Scenario #1

Review the described scenario and recommend the appropriate quality methodology to resolve the problem. Then, provide feedback on how an MES would help in the scenario.

A manufacturing floor operates a single 8-hour shift (1 hour used for breaks and lunch) five days a week and has a process takt time of one unit every minute (total production capacity of $1 \bullet 60 \bullet 7 = 420$ units per day). The production line is dedicated to the specific product with each operation having an average cycle time of 40 to 50 seconds and one operation (testing) that takes 58 seconds (approximately 1 minute). The current first-pass yield is 70%, giving a first-pass yield throughput of 294 units per day.

The manufacturing manager has been given a requirement to increase the production line output to 357 units per day. Which quality methodology would provide the best results, and how can MES help?

7.3 CI Scenario #2

Detail the MES processes and data required to support improving the following scenario:

The line-side-storage part bin for a particular workstation is emptied at least four times per shift, causing a 15-minute work stoppage each time while the replacement bin is found and delivered to the line-side stocking (LSS) location. Upon investigation, it is determined that the people who bring the parts to the line side are not notified that this bin requires replenishment, whereas there is a notification system in place for other part bins in the plant.

8 New Product Introduction: Details and Benefits

8.1 A Successful New Product Introduction (NPI)

Select the MES functions that can help support a successful NPI.

- ☐ By tracking material batch IDs during production, an MES provides insight into material performance for vendor management and material selection.
- ☐ The MES helps forecast and track the estimated material costs throughout the production process and through repair or rework.
- ☐ The MES tracks multiple small batches for product or testing qualification.
- ☐ The MES provides material use tracking for an inventory audit.
- ☐ The MES provides standard work instructions for production operators.

8.2 NPI Prototype Support

True or False: The first requirement of using an MES in an NPI environment is to stabilize the process.

Why? ____

____

____

____

____

____

Why is an MES of value when used in an NPI environment?

____

____

____

____

____

____

9 MES Architectural Concepts

9.1 ISA-95 Levels

Align each ISA-95 level with the resource.

Level		Resources
Level 0	______	A. Fabrication activity
Level 1	______	B. MES or LIMS
Level 2	______	C. Devices (sensors, actuators)
Level 3	______	D. ERP, PLM systems
Level 4	______	E. PLC/SCADA

9.2 Industry x.0 Classifications

Match the development to the Industry x.0 classification.

Classification		Development
Industry 1.0	______	A. Computation and IT at device levels using large amounts of data
Industry 2.0	______	B. PLCs
Industry 3.0	______	C. Conveyance systems
Industry 4.0	______	D. Power (water, steam, and eventually electricity) to drive large-scale manufacturing processes

9.3 Introducing MES

In many manufacturing companies that have installed ERP and PLM systems, the applications are linked using a dedicated interface. Comment on the problems that can arise when introducing MES. How are these problems best rectified?

9.4 Challenges to MES Architecture

Select the challenges that apply directly to MES architecture. Not all the items listed apply; select only those that do.

- User experience
- Common standard data models
- Network security
- IIoT devices
- Database complexity
- WAN data transfer latency

9.5 Database Structures

Two database structures are used in an MES system: star schemas and data warehouses. What type of reporting is best suited for each of these database structures?

10 Differences in Manufacturing Models and Their Impact on MES

10.1 Manufacturing Types

Match the manufacturing type to its description.

Manufacturing Type		Description
Discrete	________	A. The production process involves continuous processing of material to produce large batches of the end product.
Batch	________	B. More than one manufacturing type is used to manufacture the end product.
Continuous	________	C. Items produced are tracked in lots that contain multiple items.
Hybrid	________	D. Each item produced is individual and unique.

10.2 Manufacturing Specifics

Describe each type of manufacturing.

Automotive

____

____

Aerospace

Food and Beverage

Metallurgy and Polymers

Pharmaceutical

10.3 MES Functional Support by Industry

Fill in the MES Function versus Industry Table.

MES Function versus Industry

Manufacturing Item	Automotive	Aerospace	Consumer Goods	Electrical/ Electronic	Food & Beverage	Metals & Plastics	Pharmaceutical
Type							
Discrete							
Batch							
Continuous							
Hybrid							
Order Execution							
Just-in-Time Scheduling							
Order Launch Broadcast							
Order Tracking							
Shipping							
Material Management							
E-kanban							
Kitting							
Line-Side Replenishment							
Subassemblies							
Quality & Traceability							
Error Proof/Poka-Yoke							
Test Plans							
Trace/Genealogy							
Work Instructions							
Production Monitoring							
Equipment Alarms							
Production Alarms							
OEE Metrics							
Throughput							

11 Developing an MES Strategy

11.1 Best Practice in MES Implementation

What is the best practice in project ownership for MES and why?

____

____

____

____

____

11.2 Develop an MES Strategy

The following list contains the steps for developing an MES strategy. Place them in sequential order.

1. ______	A. Evaluate COTS software offerings to documented processes; ensure there is a way for a continuous improvement business model.
2. ______	B. Choose the solution.
3. ______	C. Determine the model of manufacturing (discrete, batch, continuous).
4. ______	D. Determine the type of manufacturing (automotive, aerospace, pharmaceutical, etc.).
5. ______	E. Develop and deploy the solution.

6. ________ F. Evaluate the deployment.
7. ________ G. Monitor usage of the system for process and system improvements.
8. ________ H. Determine and review any manufacturing regulations.
9. ________ I. Determine costs for developing and supporting in-house software.
10. ________ J. Document processes for order management (fulfillment), quality assurance, material handling, and production monitoring.
11. ________ K. Compare COTS with an in-house solution.

11.3 MES Implementation Teams

Identify the three main methods used by MES implementation teams and the pros and cons of each.

12 Current Manufacturing Challenges and Trends

12.1 Brownfield Plants and Aging Assets

Select the concepts that match the challenges of brownfield plants and aging assets.

- ☐ Equipment or other assets with minimal external communications ability
- ☐ Processes that were missed during the design phase of the plant
- ☐ Making major changes to processes without affecting production schedules
- ☐ Lack of consistency in performing a business function across multiple plants
- ☐ The cost of upgrades and training

What action can be taken to minimize the effects of these challenges?

____

____

____

12.2 Greenfield Plants

Select the concepts that match the challenges of greenfield plants.

- ☐ Processes that were missed during the design phase of the plant
- ☐ Equipment of other assets with minimal external communications ability
- ☐ Ad hoc solution development that does not fit into overall process modeling
- ☐ Lack of best practices developed to support the business

What action can be taken to minimize the effects of these challenges?

____

____

____

____

12.3 Implementing Standards

Identify three levels of standards that may impact a company's process definition.

1. ______________________________________
2. ______________________________________
3. ______________________________________

Identify three areas of an SOP that can be standardized to improve the training and effectiveness of a production operator.

1. ______________________________________
2. ______________________________________
3. ______________________________________

13 Bringing It All Together and Selecting a Solution

13.1 Effects of Variation

What is the primary concern of "common cause variation," and how does an MES help?

____

____

____

____

____

13.2 Impact on Sales Forecasting

True or False: Operations should monitor the consumption of the sale forecast.

Why? ____

____

____

____

____

13.3 Control of Continuous Improvement Initiatives

How is a continuous improvement initiative like an investment?

____

____

____

____

____

____

In what two ways does MES help manage the continuous improvement *investment*?

1. ____

   ____

   ____

2. ____

   ____

   ____

13.4 Requirements

Select the items that would make up a good MES requirements document. Some of the items listed may not be applicable.

- ☐ Simulation
- ☐ Business/manufacturing processes directly supported
- ☐ System architecture (cloud, plant floor devices, etc.)
- ☐ Security
- ☐ Data model architecture/structure
- ☐ Industrial work instructions – product or process
- ☐ Hardware: scanners and printers
- ☐ Networking
- ☐ Features and functions

- ☐ Dependencies on other systems, like ERP and PLM
- ☐ Dashboards
- ☐ Reports
- ☐ Regulations
- ☐ Data retention

Part II

Answer Key

1 Introduction, Overview, and the Current State of Manufacturing

1.1 What is ERP?

What does ERP stand for, and what are the high-level processes?

ERP stands for *enterprise resource planning* and is the set of processes that includes business and capacity planning, scheduling, and customer order management.

1.2 What is PLM?

What does PLM stand for, and what are the high-level processes?

PLM stands for *product lifecycle management* and is the support for business processes that manage the whole process of design, engineering, manufacturing planning, services, and the eventual decommissioning of a product (or product line).

1.3 What is MES?

What does MES stand for, and what are the high-level processes?

MES stands for *manufacturing execution system* and includes the detailed processes of both business and manufacturing used to produce a product, as well as the use of technology to ensure that the product is manufactured correctly. It also is used to ensure the equipment to be used during manufacturing is in good working condition to protect product quality.

1.4 What is MOM?

What is MOM, and what are the four main areas of operations that MOM covers?

MOM stands for *manufacturing operations management*, and it defines the modeling of processes used to plan, deliver, and support product manufacturing. The four main areas of operations that MOM covers are production management, quality management, inventory management, and production maintenance management.

1.5 Manufacturing Enterprise Planning

What are the five main areas of manufacturing enterprise planning within the plant?

1. Facility planning and readiness
2. Business and manufacturing process creation and development
3. Tooling and conveyance hardware readiness
4. Plant floor software development and readiness
5. ERP, MES, and PLM readiness

1.6 ISA-95 Level Order

Assign the activities to the appropriate ISA-95 levels.

Level	**Activities**
Level 0	Production activity
Level 1	Sensors and devices that support production
Level 2	Control and monitoring
Level 3	Manufacturing operations management
Level 4	Business planning and logistics

2 A Functional Overview of MES

2.1 Level 3 and Level 4 General Interfaces

Identify the general blocks of data that are included in the interfaces between Level 3 and Level 4.

- Material movement to and from production
- Production order and order reply confirmation
- Quality plan and quality results information
- Production and operator tracking and control
- Machine/process status tracking and management
- Operator work instructions

2.2 Level 2 and Level 3 General Interfaces

Identify the general blocks of data that are included in the interfaces between Level 2 and Level 3.

- Product movement management
- Equipment staging and status monitoring
- Equipment user transactions

2.3 Difference between ERP and MES

What is the primary difference between ERP transaction processing and MES transaction processing? Why is this difference significant?

> ERP uses transaction queuing to manage the execution of transactions and, therefore, cannot provide real-time feedback to production operators (i.e., operators are not required to wait for the current transaction to complete). MES requires real-time processing of transactions and therefore requires the operator to wait for the transaction to complete before moving to the next transaction.

2.4 MES Functionality

List and briefly describe the three primary categories of MES functionality.

1. **Basic track and trace** – Controlling and monitoring product movement through the production process and tracking product genealogy
2. **Facilities management and process support** – Controlling and monitoring equipment status and real-time reporting of operations activity
3. **Advanced simulation and predictive services** – Using historical production information to simulate production throughput and monitor production entities to predict the likelihood of failure

3

Order Execution Management: Details and Benefits

3.1 Benefit of MES to Order Execution

Identify a primary value that MES can provide to production planners.

The MES provides detailed tracking of production order status in real time.

3.2 Other Benefits of MES

Identify other general areas of interest and explain the value that MES provides in the context of order execution.

- **Basic track and trace** – Because MES records contain detailed data about all production-level transactions, MES can provide reports to various departments within Operations by converting the data collected into different formats.
- **Getting an early warning** – Using the mBOM and production routing, MES can determine whether a production order is on schedule for completion and show the impact on other production orders using free-running dashboards.

- **Maintain cycle-time reference** – By recording the actual cycle times of production movement, MES can provide information on changes to standard cycle times for use in production planning.
- **Labor tracking** – Because the MES records the touch time of all production operators, it can provide information on changes to labor costs for use in production planning.
- **Shipping and advanced ship notice** – By using MES in container management (similar to genealogy management of a product), the MES can record the serial numbers of products placed on a skid to create the bill of lading and an advanced ship notice. These documents can be sent to a customer's Receiving department to speed up inventory receipt at the customer level.

4 Material Management: Details and Benefits

4.1 Material/Part Lifecycle

Place the lifecycle stages of the material or part in order.

1. Plant orders the part.
2. Vendor ships the part.
3. Plant receives the part.
4. Part is transferred to the plant warehouse.
5. Parts are transferred from bulk storage to the shopping area.
6. Parts are picked in the order needed by production.
7. Parts are delivered to the production line.
8. Parts are consumed and removed from plant inventory.

4.2 Material State

Place the material states in order, and indicate which states are considered to be part of the in-plant inventory.

1. Ordered
2. Received **
3. In the warehouse **
4. In the shopping market **
5. WIP**
6. Consumed

** Part of the in-plant inventory

4.3 Material Management Methods

Describe the following material management methods:

- **E-kanban** – *Kanban* refers to a process where parts inventory at the line is monitored at a small batch level called a kanban container. When a kanban container is emptied (resulting in a kanban card being open), a manual process is triggered to replenish it with a defined amount of specific parts. *E-kanban* refers to using IT methods to automatically trigger the replenishment process and provide a *container tracking* method that is managed by the IT method.
- **Kitting** – *Kitting* refers to putting together a defined set of parts. Kitting can refer to two types:
    1. Product-specific kitting
    2. Production sequence kitting
- **Subassemblies** – A subassembly is a set of parts that are assembled before being sent to the main line for assembly on the final product. With material management, subassembly parts can be more closely controlled with an initial part pick for the subassembly and then with the actual part pick at the subassembly stations.
- **Bulk parts** – Bulk parts are universally used over the entire range of end products and are not tied to option content. These parts may include screws and

other metal and plastic fastening objects or parts that do not require special handling containers. Because these parts are not linked to option content and therefore may not be on every end product, the consumption is not tracked by part pick lists and only needs to be tracked by the number of parts used on the end product. Once this consumption rate is determined, and the quantity of inventory required in the plant is determined, ordering replenishment parts from the vendor can be automated.

4.4 Automated MES Material Management

What effect does using an automated MES material management function have on inventory management in general?

An automated MES material management function helps ensure material inventory is more accurate and provides automatic reordering to protect against stock-outs and too much inventory.

4.5 Selecting the Proper Benefits

Select the benefits that an MES provides for material management.

The following are benefits:

- It supports just-in-time production scheduling.
- It minimizes in-plant inventory, which can be a taxable commodity.
- Inventory is more accurate.
- Inventory is leveled to maximize production while minimizing inventory.
- It directly supports the kitting process by automating the replenishment notification. It also helps minimize selecting incorrect parts and causing end-product quality issues.
- It streamlines the subassembly process to minimize defects before the final assembly process.
- It automates the ordering functionality and in-plant transfers. Ordering and transfers no longer require a manual process that may not be triggered at the proper time.

- It provides the ability to track parts and products through the manufacturing lifecycle.
- It supports real-time fault annunciation of all different material management faults that negatively affect production. This speeds up resolution time.
- It automatically backflushes parts, which in turn triggers payment. Vendor payments no longer require a manual process that may not be triggered at the proper time.
- It ensures that parts are sequenced properly when picked from the shopping area.

The following are not material management benefits:

- It minimizes scrap. This is a quality issue.
- It helps ensure proper part picking at the production line side. Line-side part pick assurance is a quality function.
- It annunciates equipment faults. This is a production monitoring function.
- It helps define the order schedule. This is an order fulfillment function.

5 Quality and Traceability: Details and Benefits

5.1 Quality Control versus Quality Assurance

Explain the difference between quality control and quality assurance.

Quality control answers the question, "Was it built right?" after the process (or subprocess) is complete. Quality assurance answers the question, "Is the process understood enough to ensure the product will be built right?" before the process (or subprocess) starts.

5.2 What is SPC?

What does SPC stand for? How does it support quality assurance in an MES context?

Statistical process control. In manufacturing, SPC represents a sampling of products and evaluation of the quality metrics in order to determine if the manufacturing process is getting out of control and requires intervention.

5.3 Poka-yoke

Identify two of the poka-yoke methods used in MES.

1. Process data sequence and data format control to ensure the correct sequence of steps and accurate data entry
2. Pick-to-light to ensure that the correct part is selected for assembly

5.4 MES Traceability

List four aspects of manufacturing where the traceability provided by MES can be of value to a manufacturing company, and briefly explain why it is of value.

1. **Traceability and process improvement** – The MES can collect and process a significant amount of data and link that data to events that may be the root cause of process issues; this makes it easier to isolate problems for continuous improvement.
2. **Traceability in quality control** – This can be used to ensure that the correct material went into the product and the right process was followed.
3. **Supplier management** – MES can provide accurate information on the usage and handling of material and provide feedback on material performance. This is used to highlight suppliers whose material is performing well or not.
4. **Warranty management and customer service** – MES can provide accurate genealogy tracking to make it easier to handle recalls and identify customers that may be affected by a recall.

Why is traceability an important aspect of quality?

Traceability is an important aspect of quality because pertinent data that is collected throughout the manufacturing process can be used to support troubleshooting activities when issues arise. By reviewing the process data, the troubleshooting team can determine if an error occurred in the manufacturing process and if other units were affected by the same issue.

6 Production Monitoring: Details and Benefits

6.1 Fault Type and Priority

Classify the following faults by type and priority. Fault types include Equipment, Production, and Quality. Priorities include Critical, Major, and Minor. Assume there are no redundant systems used in this example.

Note: Below is just one example of fault classification. Fault classifications can vary from plant to plant and from manufacturer to manufacturer. Fault classifications are usually a manufacturing operations management (MOM) decision made by the manufacturing engineers. This is just an MES configuration example. This book is not intended to drive any potential MOM standards. Please refer to the MOM standards in ISA-95 for specific information.

Fault	Description	Type	Priority
Workstation A1 main power off	Main power is off at workstation	Equipment	Critical
Lift station X3 motor overload	Motor is overloaded and not working	Equipment	Critical
Buffer zone C10 full	Buffer zone near column C10 is full	Production	Major
Workstation C3 overcycle	Workstation C3 has had an overcycle	Production	Minor

Fault	Description	Type	Priority
Workstation E50 position sensor needs calibrating	Sensor needs calibration	Equipment	Minor
Workstation H5 part bin 2 is empty	Part bin containing the subpart for assembly is empty	Production	Major
Workstation H5 part bin 3 is full	Part bin containing the subpart for assembly is full and cannot accept any more parts	Production	Minor
Line A2B temperature sensor 1 needs calibration	Check calibration	Equipment	Minor
Line B2C proximity sensor is faulted	Sensor checks for proper part alignment	Equipment	Major
Workstation XYZ overcycle	Cycle time exceeded	Production	Minor
Workstation V1A misassembly	Assembly process in the station was incorrect	Quality	Major/ Minor
Inspection station Q2A: part failed stress test	Material failure	Quality	Minor
Inspection station FQ part failed operational test	Part did not operate as per spec	Quality	Minor

6.2 Equipment Trend Case Study

Review the equipment trend study of the top 10 equipment faults. Pick the top three and explain the reasons for your selections.

Top 10 Equipment Faults Report for One Month

Fault	Description	# of Occurrences	Average Downtime	Total Downtime	Replacement Effort
Motor overload	Motor stopped due to overload	10	8	80	Replace motor control relay Time: 45 minutes Cost: $250 per relay
Temp sensor fault	Sensor stopped working	12	5	60	Replace sensor Time: 20 minutes Cost: $500
Pump fault	Pump is not producing flow	2	30	60	Replace pump Time: 30 minutes Cost: $220
Transfer gate stuck open	Part gate stuck open	25	2	50	Replace gate actuator Time: 35 minutes Cost: $120

Fault	Description	# of Occurrences	Average-Downtime	Total-Downtime	Replacement Effort
Transfer gate stuck closed	Part gate does not open	20	2	40	Replace gate actuator, Time: 35 minutes Cost: $120
Safety barrier breached	When breached, equipment shuts down for safety	20	1.5	30	Replace light screen if necessary Time: 15 minutes Cost: $500
Actuator failed to move	Electrical actuator	1	25	25	Replace actuator Time: 20 minutes Cost: $400
Press fault	Press failed to achieve pressure	2	12	24	Tighten connectors, replace hoses Time: 15 minutes per hose Cost: $50 per hose
Hydraulic pressure fault	Failed to achieve starting pressure	2	11	22	Tighten connectors, replace hoses Time: 15 minutes per hose Cost: $50 per hose
Conveyor B4 drive chain fault	Chain from the drive gear to conveyor broken	1	20	20	Replace chains Time: 20 minutes per chain Cost: $80 per chain
Note: Downtime is in minutes.					

Reason table

The list should be prioritized based on the highest downtime impact with the lowest cost and repair time.

Item	Fault	Reason
1	Transfer gate	Total downtime for both faults: 90 minutes Repair time: 35 minutes Cost: $120
2	Motor overload	Downtime: 80 minutes Repair time: 45 minutes Cost $250
3	Pump fault	Downtime: 60 minutes Repair time: 30 minutes vCost: $220

6.3 Cycle Time for Improvement

Make three suggestions to improve productivity based on the cycle time trend for a manufacturing line that is serial: Station 1 directly feeds Station 2, which feeds Station 3, which feeds Station 4, which feeds Station 5. There are no parallel operations. After an assembly has entered Station 1, it cannot come out until Station 5. The line must produce one part per minute, so the cycle time for each station must be 60 seconds or less.

Cycle	1	2	3	4	5
1	59.5	50.0	61.2	56.0	59.5
2	59.0	45.9	65.0	56.0	59.0
3	58.0	51.5	64.8	56.0	58.0
4	59.9	56.0	64.9	56.1	59.9
5	60.0	47.8	65.2	56.1	60.0
6	58.5	46.7	65.1	56.0	58.5
7	59.0	49.2	66.0	56.1	59.0
8	59.3	53.3	64.4	56.2	59.3
9	59.5	54.4	64.9	56.1	59.5
10	59.6	55.5	64.5	56.0	59.6

1. Station 3 is consistently overcycling (cycle time is greater than the 60-second requirement). This should be investigated to determine whether the operator needs further training, the work there needs to be streamlined, or the subparts are not ready.
2. Every cycle time for Station 2 is well under 60 seconds, but the time is inconsistent. We need to find out what is going on and stabilize the process.
3. Determine if some work from Station 3 can move to Station 2 to balance the line.
4. Station 1 and Station 5 are currently within cycle time requirements; however, if Station 3 is corrected, either Station 1 or Station 5 may become the next constraining operations.

6.4 OEE Case Study for Improvement

The following is an OEE report for a line for one shift. Calculate the return on investment (ROI) to raise both the production performance and the first-pass quality rate by

0.05 at an investment cost of $50,000 if each part produced generates revenue of $2000 and costs an average of $1200 per part to repair before shipping.

OEE Calculation per Hour Shift = 8 hours									
Scheduled JPH		60							
Hour	Downtime	Availability	Scheduled JPH	Actual JPH	Performance	Actual Quality	Quality Rate	OEE	Reasons
1	15	0.750	60	40	0.667	35	0.875	0.438	Equipment Down for Repair
2	0	1.000	60	50	0.833	45	0.900	0.750	
3	0	1.000	45	40	0.889	35	0.875	0.778	15 minute break
4	0	1.000	60	51	0.850	45	0.882	0.750	
5	0	1.000	30	25	0.833	22	0.880	0.733	30 minute lunch
6	0	1.000	60	51	0.850	45	0.882	0.750	
7	0	1.000	45	40	0.889	35	0.875	0.778	15 minute break
8	0	1.000	60	53	0.883	46	0.868	0.767	
Shift Total:	15	0.969	420	350	0.833	308	0.880	0.710	

Use shift totals.

To increase the production performance by 0.05, multiply 0.05 by 350 (actual JPH) to get 17.5.

To increase the first-pass quality rate by 0.05, multiply 0.05 by 308 (actual quality) to get 15.4.

To get the revenue increase for production performance, multiply $2000 by 17.5 to get $35,000.

To get the cost savings for quality, multiply $1200 by 15.4 to get $18,480.

Total cost improvement in one shift is $35,000 plus $18,480 for a total of $53,480.

The ROI on $50,000 is less than one shift.

7 Continuous Improvement: Details and Benefits

7.1 Process Improvement Steps

Place the following steps in order.

1. Stabilize the process.
2. Characterize the stable process.
3. Determine the most significant issue for the process.
4. Determine the root cause of the most significant issue.
5. Make process improvements to remove (or control) the most significant issue.
6. Evaluate the effectiveness of the change.
7. Repeat step 1.

7.2 CI Scenario #1

Review the described scenario and recommend the appropriate quality methodology to resolve the problem. Then, provide feedback on how an MES would help in the scenario.

A manufacturing floor operates a single 8-hour shift (1 hour used for breaks and lunch) five days a week and has a process takt time of one unit every minute (total production capacity of $1 \bullet 60 \bullet 7 = 420$ units per day). The production line is dedicated to the specific product with each operation having an average cycle time of between 40 to 50 seconds and one operation (testing) that takes 58 seconds (approximately 1 minute). The current first-pass yield is 70% giving a first-pass yield throughput of 294 units per day.

The manufacturing manager has been given a requirement to increase the production line output to 357 units per day. Which quality methodology would provide the best results, and how can MES help?

Using the theory of constraint, there are two options:

1. Have production maximize the testing operation during breaks and lunch, and then add overtime to achieve the required throughput to maximize the constraint. Use MES to monitor a buffer before testing and the testing throughput to ensure the production rate is being met.
2. Because only 70% of production is passing within the first-pass yield (restricting the throughput of testing), implementing a continuous improvement program based on improving the first-pass yield is recommended. Use MES to provide Pareto reports of the fallout at testing, then analyze the testing data and the process data collected to determine the cause of the 30% fallout.

7.3 CI Scenario #2

Detail the MES processes and data required to support improving the following scenario:

The line-side-storage part bin for a particular workstation is emptied at least four times per shift, causing a 15-minute work stoppage each time while the replacement bin is found and delivered to the line-side stocking (LSS) location. Upon investigation, it is determined that the people who bring the parts to the line side are not notified that this bin requires replenishment, whereas there is a notification system in place for other part bins in the plant.

This is an MES material-handling process issue.

The requirements will be to monitor the material consumption rate from the LSS bin and create a notification to the ERP system when a replenishment is required.

The required MES data points are:

- Current part count in the line-side storage bin
- Max count set point for the storage bin (What is the maximum number of parts for the bin?)
- Reorder set point (Once the current part count drops below this set point, a part reorder is issued.)
- Part-bin-low fault and set point

MES functional description:

- Define an LSS location within the MES system for the material to be consumed at this operation.
- Interface the MES to the ERP system for material replenishment (configuration varies according to the MES implemented).
- For this LSS location, set the material part number and order quantity (determined by the max count), and define the minimum quantity of material (reorder set point).
- Within the operator's dashboard, establish material consumption from the LSS location (configuration varies according to the MES implemented). As the production operator performs the required operation, material is consumed from the LSS record within the MES.
- When the available material at the LSS location reaches the minimum quantity for reorder, the MES sends a reorder notification to the ERP system.
- The LSS location quantity is updated in the MES when the material arrives.

8 New Product Introduction: Details and Benefits

8.1 A Successful New Product Introduction (NPI)

Select the MES functions that can help support a successful NPI.

The following are used to help support a successful NPI:

- By tracking material batch IDs during production, an MES provides insight into material performance for vendor management and material selection.
- The MES tracks multiple small batches for product or testing qualification.
- The MES provides standard work instructions for production operators.

The following are not:

- The MES helps forecast and track the estimated material costs throughout the production process and through repair or rework.
    - The ability to help forecast and track material costs is a function of normal production and is not a priority in NPI.

- The MES provides material use tracking for an inventory audit.
  - Inventory management is not a priority in the management of NPI.

8.2 NPI Prototype Support

True or False: The first requirement of using an MES in an NPI environment is to stabilize the process.

Why?

False. One of the primary purposes of NPI is to develop manufacturing processes. Within the NPI environment, multiple versions of a product and product build levels may be active at any time. As a result, the concept of a "stable process" may not be possible.

Why is an MES of value when used in an NPI environment?

An MES can track each individual product moving through manufacturing and provide detailed tracking and control data for all material used for each product version. This capability ensures that manufacturing and product engineers have the correct data to use to evaluate products and processes.

9 MES Architectural Concepts

9.1 ISA-95 Levels

Align each ISA-95 level with the resource.

Level	**Resources**
Level 0	Fabrication activity
Level 1	Devices (sensors, actuators)
Level 2	PLC/SCADA
Level 3	MES or LIMS
Level 4	ERP, PLM systems

Note: As IIoT and cloud-based technologies continue to evolve, hardware in Levels 1 through 4 may support multiple levels. For example, as plant floor devices become more capable, they may end up supporting Levels 1 through 3. Also, as cloud-based systems become more stable and robust, they may support Levels 2 through 4 in.the future.

9.2 Industry x.0 Classifications

Match the development to the Industry x.0 classification.

Classification	Development
Industry 1.0	Power (water, steam, and eventually electricity) to drive large-scale manufacturing processes
Industry 2.0	Conveyance systems
Industry 3.0	PLCs
Industry 4.0	Computation and IT at device levels using large amounts of data

9.3 Introducing MES

In many manufacturing companies that have installed ERP and PLM systems, the applications are linked using a dedicated interface. Comment on the problems that can arise when introducing MES. How are these problems best rectified?

The traditional interface between an ERP system and a PLM is a dedicated link between the applications. When trying to implement an MES into this system, implementers are frequently required to use the same interface to transfer data from the PLM to the MES. This forces the ERP system to handle data that it has no reference for and creates a lot of customizations that are costly to maintain. To rectify this problem, implementers should use SOA as the link between the ERP system and the PLM and from either system to the MES.

9.4 Challenges to MES Architecture

Select the challenges that apply directly to MES architecture. Not all the items listed apply; select only those that do.

The following are challenges to MES architecture:

- **Common standard data models** make it easier to transfer data and are the backbone of autonomous MES. The challenge is creating the common data model, which can be a large project. Too often there is no common data model.
- **Database complexity** affects the ability to interface data between systems.
- **IIoT devices** add complexity to the manufacturing process and must be carefully planned for.
- **WAN data transfer latency** determines which operations can move to the cloud and which operations must remain on-site.

The following are not:

- **User experience** is affected by many things, not just system architecture responsiveness.
- **Network security** is its own discipline and affects many things including architecture and IIoT devices.

9.5 Database Structures

Two database structures are used in an MES system: star schemas and data warehouses. What type of reporting is best suited for each of these database structures?

The star schema is used for operational reporting for operations entities' current status and production event notification. The data warehouse is best used to generate trend-type reports.

10 Differences in Manufacturing Models and Their Impact on MES

10.1 Manufacturing Types

Match the manufacturing type to its description.

Manufacturing Type	Description
Discrete	Each item produced is individual and unique.
Batch	Items produced are tracked in lots that contain multiple items.
Continuous	The production process involves continuous processing of material to produce large batches of the end product.
Hybrid	More than one manufacturing type is used to manufacture the end product.

10.2 Manufacturing Specifics

Describe each type of manufacturing.

Automotive

- This type is high-volume, medium-option content manufacturing.
- Scheduling is detailed to a Just-in-time methodology.
- Metalfab, body shop, powertrain, and parts use both batch and discrete manufacturing.
- Paint shop uses batch manufacturing.
- General assembly uses only discrete manufacturing.

Aerospace

- This type involves low-volume, high-option content.
- Batch processes are used for parts and discrete processes are used for assemblies throughout the manufacturing lifecycle.
- Work must be certified.
- Regulations are global and tend to be fairly standard.
- Quality assurance is assessed through traceability and genealogy.

Food and Beverage

- Quality assurance is key, focused on the proper scheduling and tracking of maintenance, cleaning, and upkeep in order to minimize contamination.
- Batch process parameters are closely monitored for traceability and genealogy requirements.
- For safety, product and processes are highly regulated.

Metallurgy and Polymers

- Materials are processed to eliminate impurities.
- Continuous manufacturing is used.

- Production monitoring to maintain continuous manufacturing and quality assurance is key.

Pharmaceutical

- Quality assurance is key, starting with the purity of the materials.
- This type focuses on properly scheduling and tracking maintenance, cleaning, and upkeep to minimize contamination.
- Processes are highly regulated, and clearly defined reports are required.

10.3 MES Functional Support by Industry

Fill in the MES Function versus Industry Table.

MES Function versus Industry

Manufacturing Item	Automotive	Aerospace	Consumer Goods	Electrical/ Electronic	Food & Beverage	Metals & Plastics	Pharmaceutical
Type							
Discrete	X	X		X			
Batch			X	X	X		X
Continuous						X	
Hybrid	X	X		X			
Order Execution							
Just-in-Time Scheduling	X	X					
Order Launch Broadcast	X						
Order Tracking	X	X		X			
Shipping	X		X	X	X	X	X
Material Management							
E-kanban	X			X			
Kitting	X	X		X			
Line-Side Replenishment	X	X	X	X	X	X	X
Subassemblies	X	X	X	X			

Manufacturing Item	Automotive	Aerospace	Consumer Goods	Electrical/ Electronic	Food & Beverage	Metals & Plastics	Pharmaceutical
Quality & Traceability							
Error Proof/Poka-Yoke	X	X	X	X			
Test Plans			X	X	X	X	X
Trace/Genealogy	X	X	X	X	X	X	X
Work Instructions	X	X		X			
Production Monitoring							
Equipment Alarms	X	X	X	X	X	X	X
Production Alarms	X	X	X	X	X	X	X
OEE Metrics	X		X	X	X	X	
Throughput	X		X	X	X		X

11 Developing an MES Strategy

11.1 Best Practice in MES Implementation

What is the best practice in project ownership for MES and why?

> The best practice for MES implementation projects is to have all operational IT systems and OT technology managed (and funded) under Operations control. This is because Operations is the primary (and sometimes only) benefactor of the implementation, and it provides common management for both IT and OT initiatives.

11.2 Develop an MES Strategy

The following list contains the steps for developing an MES strategy. Place them in sequential order.

1. Determine the type of manufacturing (automotive, aerospace, pharmaceutical, etc.).
2. Determine the model of manufacturing (discrete, batch, continuous).
3. Determine and review any manufacturing regulations.
4. Document processes for order management (fulfillment), quality assurance, material handling, and production monitoring.
5. Evaluate COTS software offering to the list of documented processes; ensure that there is a way for a continuous improvement business model.

6. Determine costs for developing and supporting in-house software.
7. Compare COTS with an in-house solution.
8. Choose the solution.
9. Develop and deploy the solution.
10. Evaluate the deployment.
11. Monitor usage of the system for process and system improvements.

11.3 MES Implementation Teams

Identify the three main methods used by MES implementation teams and the pros and cons of each.

- **Turnkey Consulting**

  Pros

  - The team is responsible for developing all delivery requirements for the implementation.
  - It delivers a significant amount of functionality at the start.
  - There is a planned fixed length of engagement for the implementation.

  Cons

  - Considerable capital is needed upfront for project funding.
  - Requirements must be defined, understood and accepted by the full implementation team upfront.
  - Changes to requirements can be more difficult to approve.

- **Mentorship Consulting**

  Pros

  - Capital funding is spread over a longer period (better for cash flow).
  - Implementation provides functionality as needed.
  - The company gains a deeper knowledge of the use of the MES.

Cons

- The length of the implementation is less predictable.
- Project liability stays with the company.

- **Internal Implementation Team**

  Pros

  - Team funding is purely operational. (CapEx is for computing resources only.)
  - The company controls all the project resources and, therefore, the direction of the implementation.

  Cons

  - The company must hire all required skills internally.
  - Finding (and holding) required skills may be difficult.

12 Current Manufacturing Challenges and Trends

12.1 Brownfield Plants and Aging Assets

Select the concepts that match the challenges of brownfield plants and aging assets.

- ✓ Equipment or other assets with minimal external communications ability
- ☐ Processes that were missed during the design phase of the plant
- ✓ Making major changes to processes without affecting production schedules
- ☐ Lack of consistency in performing a business function across multiple plants
- ✓ The cost of upgrades and training

What action can be taken to minimize the effects of these challenges?

Continually planning for equipment upgrades or replacement is key to managing the cost of brownfield maintenance and aging assets.

12.2 Greenfield Plants

Select the concepts that match the challenges of greenfield plants.

✓ Processes that were missed during the design phase of the plant

☐ Equipment of other assets with minimal external communications ability

✓ Ad hoc solution development that does not fit into overall process modeling

✓ Lack of best practices developed to support the business

What action can be taken to minimize the effects of these challenges?

> Using the ISA-95 standard as a template will help to ensure that processes are not missed. Consulting information from industry organizations like ISA and MESA will support using industry standards.

12.3 Implementing Standards

Identify three levels of standards that may impact a company's process definition.

1. Regulatory
2. Industrial
3. Company

Identify three areas of an SOP that can be standardized to improve the training and effectiveness of a production operator.

1. Terminology
2. Presentation of information
3. Collection of data

13 Bringing It All Together and Selecting a Solution

13.1 Effects of Variation

What is the primary concern of "common cause variation," and how does an MES help?

> The main concern of common cause variation is that it makes problem analysis more difficult by creating noise in the analysis data.
>
> MES can help with this by reducing variation within a process, thereby reducing the noise in the process and making it easier to analyze *special cause* problems.

13.2 Impact on Sales Forecasting

True or False: Operations should monitor the consumption of the sale forecast.

Why?

> True. By monitoring the consumption of the sales forecast, Operations can gain advanced warning of unexpected changes in independent demand.

13.3 Control of Continuous Improvement Initiatives

How is a continuous improvement initiative like an investment?

Continuous improvement requires spending money to determine a solution that corrects a problem (or improves capability). After the improvement has been implemented, the solution must be maintained for a period of time (depending on the money spent) to recuperate the cost of the analysis and solution implementation.

In what two ways does MES help manage the continuous improvement *investment*?

1. An MES application is the one tool that has been explicitly designed to help MOM accurately monitor and actively manage the plant's capability.
2. MES applications were designed specifically to collect data from equipment, operators, and other systems at speeds that are fast enough to be meaningful and to store that data in a manner that maintains the relationship with other events occurring on the manufacturing floor.

13.4 Requirements

Select the items that would make up a good MES requirements document. Some of the items listed may not be applicable.

The following items make up a good requirements document:

- Business/manufacturing processes directly supported
- System architecture (cloud, plant floor devices, etc.)
- Data model architecture/structure
- Hardware: scanners and printers
- Features and functions
- Dependencies on other systems, like ERP and PLM
- Dashboards
- Reports
- Regulations
- Data retention

The following items are not in the requirements document:

- Networking: This is typically handled separately.
- Security: This is handled separately, especially if a cloud-based solution is being considered. If the cloud solution cannot meet the security requirements, no other action can be done.
- Industrial work instructions – Product or process: This is a configuration item for MES solutions that support this feature.
- Simulation: Simulation can help determine MES requirements and configuration, but it is not a requirement by itself.

Printed in the USA/Agawam, MA
September 17, 2026

910661.232

Manufacturing Execution Systems

An Operations Management Approach

Second Edition

AF871123

Preface

In the manufacturing industry, the main focus of operations personnel has been to make products as effectively as possible and to better understand the characteristics of their processes. This has resulted in the development of several methodologies to plan and schedule production and to continuously improve the product and the process of making it.

However, most manufacturing engineers and operations managers still rely on physically observing each operation with stopwatches and various sensors to build a characterization of their processes, operation by operation. Although these are the tried and true methods of the industry, they are cumbersome and slow to produce results. Even in the most recognized improvement methodologies of Lean and Six Sigma, many practitioners still rely on these practiced methods for understanding the details of a company's internal processes.

Although manufacturing execution systems (MES) were first introduced in the late 1990s, adoption of these applications has been slow and, in some cases, hotly resisted. In investigating why, it was discovered that it is mostly due to (1) a lack of understanding of how to apply MES to manufacturing operations and (2) a considerable amount of broken promises from MES vendors about delivering the silver bullet that will make everyone's jobs easier and provide more than enough return on investment (ROI) to make it worthwhile.

The missing link to vendor promotions is that, like any enterprise system, the installation alone does not make a company run better. It is in using the system that Operations

About the Authors

Grant Vokey

Grant Vokey is the principal consultant for Vokey Consulting. With 20 years of diverse manufacturing operations experience and an additional 15 years of integrating information technology (IT) systems into the manufacturing floor, Grant has developed a strong understanding of how manufacturing companies work and the information needed to operate at world-class levels.

Grant's experience, coupled with continuous training and 10 years as a Certified Operations Manager, has also provided him with an excellent understanding of industry best practices and best-in-class utilization of manufacturing execution systems (MES). Using this knowledge, he has been a subject matter expert for developing industry-leading MES applications/solutions, a program manager for multiple MES programs, and a lead consultant on implementations of MES in various verticals (electronics, industrial equipment, automotive manufacturing, and metal fabrication).

Grant has developed a reputation for providing sound, practical advice and direction that make a difference to his clients and the MES industry as a whole.

Thomas Seubert

Thomas Seubert is currently a project manager for an MES development project. Previously he worked as an MES consultant and an MES delivery and presales manager. Before that, he worked on several shop floor and MES projects going back to the year 2000 when General Motors/Electronic Data Systems were developing MES solutions.

Tom has more than 25 years of experience with manufacturing processes and controls, 18 years of experience with IT, and 17 years of experience with MES. He has consulted on MES solutions for the multitiered automotive market, as well as the aerospace, plastics and steel, tire, food and beverage, and pharmaceutical industries. He also helps instruct adults and youth in learning leadership skills within the Boy Scouts of America.

Tom was a keynote speaker at the Industrial Engineering and Operations Management Society International Conference in Orlando, Florida, in September 2015 and a recipient of their Distinguished Service Award the same year. He was also the recipient of the 2014 Boy Scouts of America District Award of Merit. He earned his BS in electrical engineering from Lawrence Technological University in Southfield, Michigan.

1 Introduction, Overview, and the Current State of Manufacturing

1.1 Introduction

Before we can discuss manufacturing execution systems (MES), we must examine the manufacturing enterprise and operation as a whole to understand how these systems fit into the big picture.

Let us start with the scope of manufacturing. Simply put, it is the entire process from procuring, receiving, and storing materials in the plant (or a nearby warehouse) to the fabrication (or assembly) and quality assurance of the manufactured part, which may include limited storage of intermediate-level parts or finished goods until they are needed, and possibly shipping the product to the first-level customer.

Manufacturing, which includes all activity within the plant, is illustrated in Figure 1-1.

It takes considerable coordination to properly manufacture a product. The following is a list of the high-level processes within the scope of manufacturing:

- **Material procurement** – When, how much, and from whom
- **Material receipt (including incoming inspection) and storage** – Frequency of delivery, where to place the delivery once it is in-house, and verification of material quality

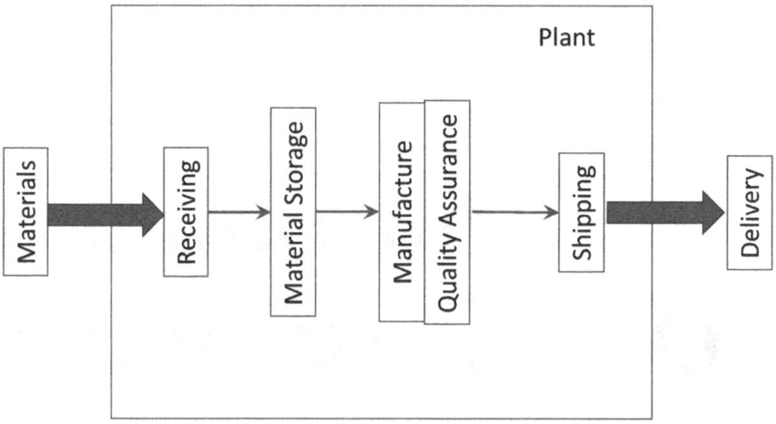

Figure 1-1. Manufacturing scope.

- **Material and product inventory management** – All material movement and storage of finished goods within the plant
- **Production planning and scheduling** – When to produce and how much, which includes sequencing the manufacturing processes for the product or subassembly at both the subassembly line level and the final assembly level
- **Maintenance and readiness of tooling and conveyance systems** – Ongoing maintenance and calibration to keep the tooling and conveyance available when needed
- **Training and readiness of operators, supervisors, and other staff** – Having both the knowledge to perform the work and the right skills available when they are needed while minimizing cost
- **Product quality assurance** – When tests should be performed on products or materials to check the product's quality while reducing potential waste if something goes wrong
- **Shipping scheduling and coordination** – Ensuring the product is shipped with enough time to deliver on time and minimizing shipping costs

These processes require the careful coordination of material, resources, and information to make and deliver the product efficiently. As such, operations must recognize that any work performed toward the fabrication or assembly of a production unit is an investment that is not recovered until the product is sold, and operations costs for producing finished goods (including scrap, direct labor, and material) must be recovered in the sales price of the product. As with any investment, Operations must try to reduce the investment needed to make each sellable product.

Beyond the full scope of manufacturing, the scope of an MES is generally from the point of *goods issue* to the production floor (material moved from inventory to *local holding* on the production floor), through the process of fabrication and/or assembly, to the point of *goods receipt* from the production floor (movement of completed products back into inventory ready for the next stage of production or sales).

All this processing is based on the current status and ongoing updates to the product design, as well as updates to valid materials used during the product's lifecycle. Additionally, demand must account for variations in the product to be delivered to the end customer as the consumer goods industry moves toward mass customization, which will potentially increase manufacturing costs. Making a product that meets customer expectations and delivering it at the lowest possible price is ultimately what manufacturing operations management (MOM) is all about. To understand how an MES affects the efficiency of producing a product, we must understand how it fits into the MOM model (see Figure 1-2).

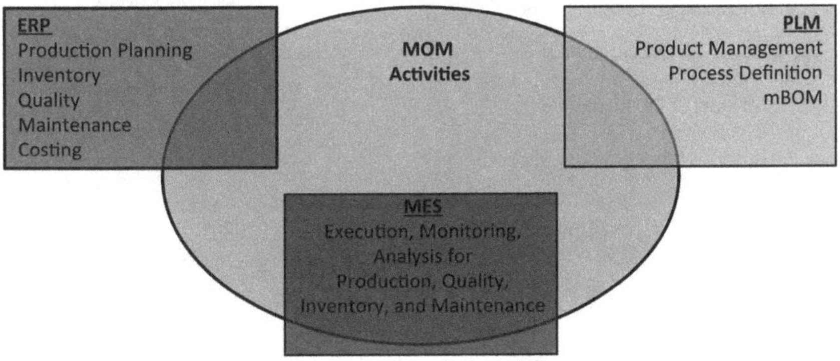

Figure 1-2. Simplified MOM diagram.

MOM (from the ISA-95 perspective) refers to the model of processes and procedures that are used to plan, deliver, and support the manufacturing of products, with most of that activity engaged in controlling and managing the production floor. This activity model manages the four main areas of operations management (production, quality, inventory, and maintenance) and includes activities for planning, monitoring, reporting, and performing an analysis (for improvement) in each of the four areas of operations as they pertain to the shop floor. The full explanation of the activity models is beyond the scope of this book, but there are some books available from the International Society of Automation (ISA) that discuss the topic further. Most of the activities that support MOM are also supported within MES. Some activity within the MOM model requires data external to MES and creates an interaction between enterprise systems such as enterprise resource planning (ERP), product lifecycle management (PLM), and MES.

1.2 ERP Business Process

ERP is the set of processes that includes business and capacity planning, scheduling, and customer order management. Accounting and Human Resources are part of these processes because they represent the management of all the resources (people, material, etc.) required to execute manufacturing. Figure 1-3 shows the business processes that are part of the ERP environment.

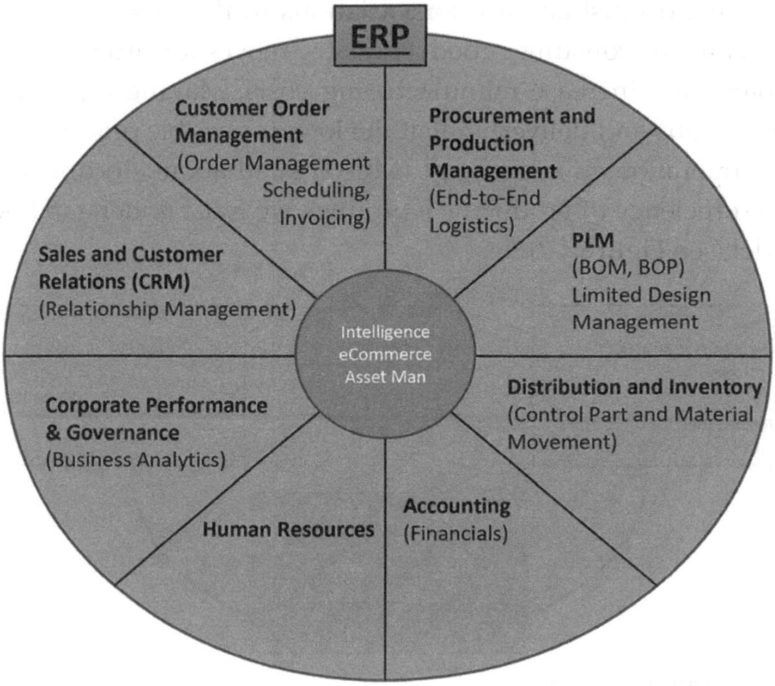

Figure 1-3. ERP modules.

A full review of the function of ERP is beyond the scope of this book.[1] For the purposes of this overview, the following functions are important for manufacturing:

- **Capacity management** – What is the maximum capacity of a manufacturing plant? This is highly dependent on the number and mix of *end products* and how they are manufactured. For every product to be manufactured, there is a defined process. There is also a maximum number of units that can be manufactured in a given time. This will affect both the production order planning and

1. For further information regarding ERP, see "Enterprise Resource Planning," Wikipedia, last modified July 14, 2019, accessed September 30, 2022, https://en.wikipedia.org/wiki/Enterprise_resource_planning.

the material procurement planning. ERP capacity management also includes functionality to analyze capacity utilization and availability.

- **Production planning** – As with everything else, manufacturing must be planned, sometimes down to the smallest element of the process (remember the efficiency of reducing the investment of time and material). In addition, as a part of production planning, there is a need for the planning staff to understand the current status of the production environment. Adjusting the production plan according to changes in this status is a major part of the production planning process. Planning must also account for changes in the current status of the production environment as changes happen.

- **Business management** – This function includes processes that provide visibility into the company's financial management (revenue, costs, and profits), inventory (raw material and finished goods), and sales management.

- **Customer order management** – These processes ensure the receipt, preparation, and planning of delivery for all customer orders, in addition to tracking order delivery and revenue collection.

1.3 Product Lifecycle Management Business Process

Product lifecycle management (PLM) is the support for the business process that manages the design, engineering, manufacturing planning, services, and eventual decommissioning of a product (or product line). Figure 1-4 shows the business processes that are part of the PLM environment.

For the purposes of this overview, the processes that are important within the manufacturing company include the following:

- **New product introduction** – The process of taking a product from concept to marketable product, which includes planning and managing necessary changes in the manufacturing environment to launch the new product. Also included, but not part of manufacturing, are the marketing and other plans to introduce the new product into the market. Because considerable investment must be made to launch a new product, it is critical to control and manage this part of the process.

- **Product design** – This will affect how the product will be manufactured. It incorporates concepts covered in a design methodology called *design for manufacturing* (DFM). Discussion of the full scope of DFM is beyond this book.

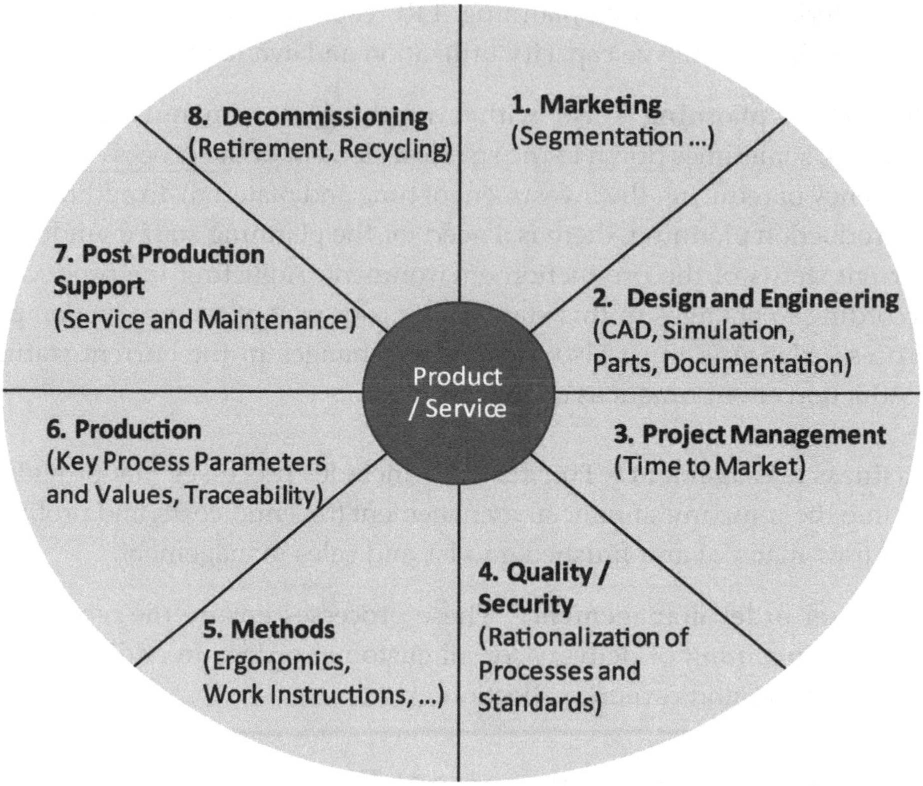

Figure 1-4. PLM model.

However, in many cases, physical limitations in the product design will have a major impact on how the product will be manufactured. This can include chemistry in material interactions, limitations in manipulating materials, and other constraints.

- **Product upgrades** – This includes changes to the product during its lifecycle. Many changes can significantly affect the manufacturing environment, from different vendors and, therefore, different supply chain management (SCM) processes to different manufacturing and quality assurance processes. Understanding when a new design should be introduced, or when a design should be upgraded during the product's lifecycle, is an important part of reducing waste caused by material obsolescence.
- **Manufacturing work instructions** – This refers to the alignment of manufacturing work instructions to design intents for assembly. As previously noted, how a product is put together is directly related to the design of the product and how it will perform after manufacturing. This takes careful coordination between Manufacturing Engineering and Product Engineering.

- **Product decommissioning** – This is the process of ramping down the product as it approaches the end of its sales life. At times, this affects how the product is manufactured as product volume decreases and required materials are depleted.

Additional information on PLM can be found on Wikipedia.[2]

1.4 MES Business Processes

An MES includes both the business and the manufacturing processes used to produce a product as well as the technology necessary to ensure that the product is properly manufactured. MES are also used to ensure the manufacturing equipment is in good working order to protect product quality. The implementation of an MES should provide maximum visibility and should control the entire manufacturing environment. In other words, if something happens on the production floor, the MES should *be aware of it*, record it, and have the capability to control it, all in real time.

An MES operates at a task level in production, and each task in the assembly process is recorded and monitored/controlled as it is happening. The meaning of *real time* in an MES is down to the time frame of fractions of seconds to record, act on, and respond to any activity that happens during the production process. In some production processes, this level of response time is not required and, as a result, an MES may not be necessary. A question that is frequently debated is, "Does an MES add value to manufacturing?" The answer is not as clear-cut as some would like it to be. An MES will always provide value to manufacturing; however, one should ask, "Does it provide enough value to justify the implementation and sustaining costs?"

Determining the answer to this question is the primary reason for this book, and the following chapters provide insight into these matters.

1.5 The Manufacturing Enterprise View

Now that we have reviewed the scope of MOM, we will explore how it fits into the multiple layers of manufacturing activity, which is best illustrated through the ISA-95 functional hierarchy model of manufacturing activities in Figure 1-5.

---

2. "Product Lifecycle," Wikipedia, last modified July 15, 2019, https://en.wikipedia.org/wiki/Product_lifecycle. Accessed October 11, 2022.

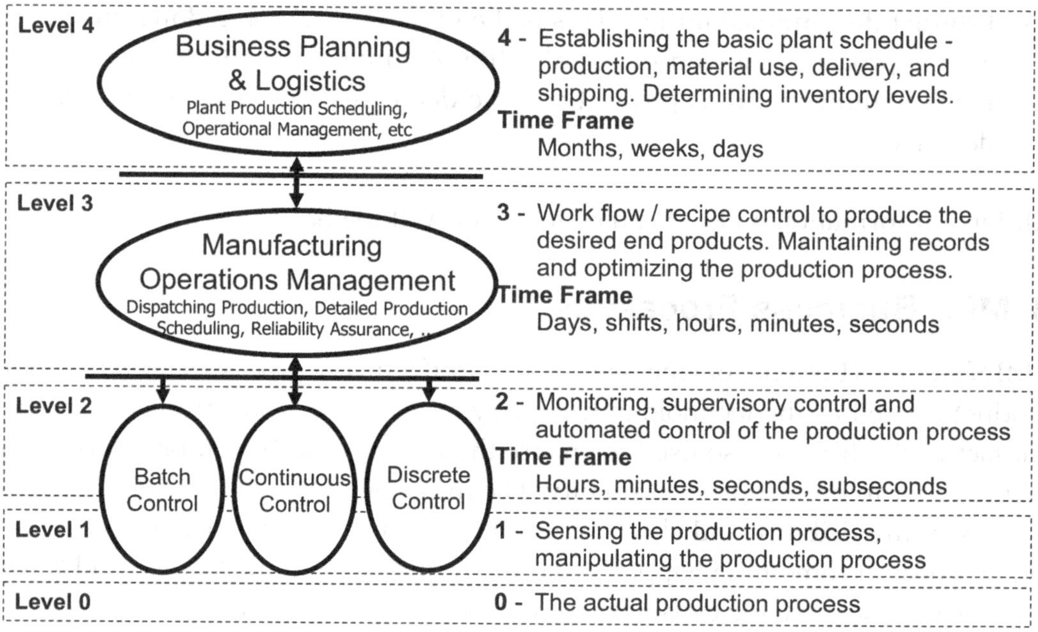

Figure 1-5. The ISA-95 functional hierarchy model. ANSI/ISA-95.00.01-2010 (IEC 62264-1 Mod), *Enterprise-Control System Integration – Part 1: Models and Terminology*, Figure 3.

Source: Reproduced with permission from the International Society of Automation (ISA).

According to the ISA-95 hierarchy model (a different perspective of the MOM activity model), the activities can be broken down as follows:

- **Level 0** – This is the actual production process. This level represents the physical hardware and people working directly on the product to complete its manufacturing process.
- **Level 1** – This represents the hardware and software (sometimes referred to as *operational technology*—OT) used to sense and manipulate the product during the manufacturing process. All sensors, conveyance motors, and other devices that act directly on the product fit into this level.
- **Level 2** – This represents the control level of the process within the scope of an operation. All systems that provide the automation and supervisory controls of the direct manufacturing process—programmable logic controllers, distributed control systems, or personal computer (PC)-based controls (supervisory control and data acquisition—SCADA)—are at this level and include functions such as limit management for Level 1 and registering event flags at the automation level to signal issues for response at Level 1. Data at Level 2 tends to have the scope of a single production line (or work center) and can be analyzed for performance at the line or operation level.

- **Level 3** – This represents the alignment of manufacturing activities to their business processes. This level includes but is not limited to production order fulfillment, quality assurance, process error proofing, and line-side subpart delivery. It also includes the data retention and reporting requirements for the *device history report* or a full history of all activity that was performed on each individual product identified by the lowest control ID available during production. The lowest level may be a batch ID or a serial number, depending on the requirements of the product and process. This level introduces manufacturing information technology (IT) and is the focus of industry work for IT and operational technology (OT) convergence. Data managed at Level 3 tends to have the scope of a single facility, including down to the line or operation level, and can be analyzed for performance at the plant level.

- **Level 4** – This represents the actual business process area. It includes but is not limited to scheduling, customer relationship management (CRM), SCM, reporting, and analytics. This model is typical of the activities within a plant for manufacturing. In manufacturing organizations that have multiple plants, the data managed at Level 4 can be analyzed for performance across the enterprise.

Within the plant, several areas must be planned for and managed:

- **Facility planning and readiness** – This ensures the facility is ready for the different elements of manufacturing. This includes but is not limited to equipment clearances; equipment placement; process, part, and product flow routes; and part/product storage and warehousing requirements. Manufacturing is usually managed within a single facility, and there are two types of facility planning that must be accounted for, each providing its own set of issues to manage:

    - **Brownfield** – The facilities already exist. The challenge here is working with existing processes, hardware, systems, and attitudes and being able to plan and manage all changes for anything new. As the changes are implemented, it is important to understand the *migration* needed to get from the prechange state to the postchange state. In some cases, planning the migration from one state to another can be just as important as planning what to change.

    - **Greenfield** – Everything, including all support infrastructures such as utilities, roads, and sometimes the building itself, is new. The challenge here is that every part of the facility is a clean slate, so details must be included early in the planning process to ensure that nothing is missed. At the same time, as the design of the facility evolves (and it will evolve), these changes must be reviewed by all aspects of the facilities design team, including all areas of manufacturing IT, to prevent the discovery of problems late in the process of commissioning the facility. Because the MES must closely mimic the production environment, reviewing changes to the facility as they occur with the MES implementation team is very important.

- **Business and manufacturing process creation and development** – All business processes—client and supplier interaction, accounting and human resources, and manufacturing planning and control—must be created, defined, and developed. Without these processes, nothing else, including systems, can be implemented. These processes must then be documented. With the implementation of most systems, process documentation is created during a project phase called *blue printing*. Significant time can be saved during the implementation phase when process documentation is initiated before system implementation begins. However, even when the documentation is initiated early, it will be updated during blue printing.
- **Tooling and conveyance hardware readiness** – This refers to all the hardware used to control and execute the manufacturing process. From the actual fabrication or assembly tooling to the part and product conveyances, all tooling and hardware must be installed and ready to operate. Tooling examples include motors, pumps, torque tools, painting equipment, mixers, welders, and any robotics. Conveyance systems include any type of conveyor (belt, power, free skids, etc.), material lifters, and material elevators or any implementation of robotic material handlers for automated inventory systems.
- **Plant-floor software development and readiness** – This refers to all software applications and systems that interact with the plant floor but are not necessarily part of the plant-floor controls solutions. All automated test equipment, quality verification software, and data recording and analysis software fit into this area, as well as supervisory systems for automated inventory management.
- **ERP, MES, and PLM readiness** – All major enterprise systems that are used to manage the ongoing activities of manufacturing must be available, including any required communications between these systems.

Planning for the readiness of the areas listed above requires many estimations based on assumptions for requirements. Part of that planning will also require planning for contingency based on scenarios that may be encountered. It is unfortunate that more companies do not engage in plant or line simulation because the process of setting up and running simulations can highlight many issues with the state of planning and can provide considerable confidence in the plan itself and in the contingency and mitigation requirements. The key to good simulation is ensuring that the correct process and equipment characteristics have been defined up front. Having prior access to data via systems like MES can go a long way to aid in developing those characteristics. As such, simulation can play a significant role in planning the initial requirements for readiness and in preparing for contingency with the simulation of capability through various what-if scenarios well in advance of investing time and money into the actual implementation of plant-level resources. Related to planning resources and processes for supporting the plant, additional effort is required to plan for the availability of operational data and information within the context of the plant. This will require additional planning for the enterprise systems (e.g., ERP and PLM) and interfaces down through MES to the shop floor systems with controls engineering and test and inspection systems.

As shown in Figure 1-6, there is a cycle around planning, simulation, and implementation. When using simulation, it is far better to have iterations between planning and simulation than to have iterations of the complete cycle. As part of the normal continuous improvement, there will always be implementation and production runtime cycles. Simulation can help minimize these cycles as well.

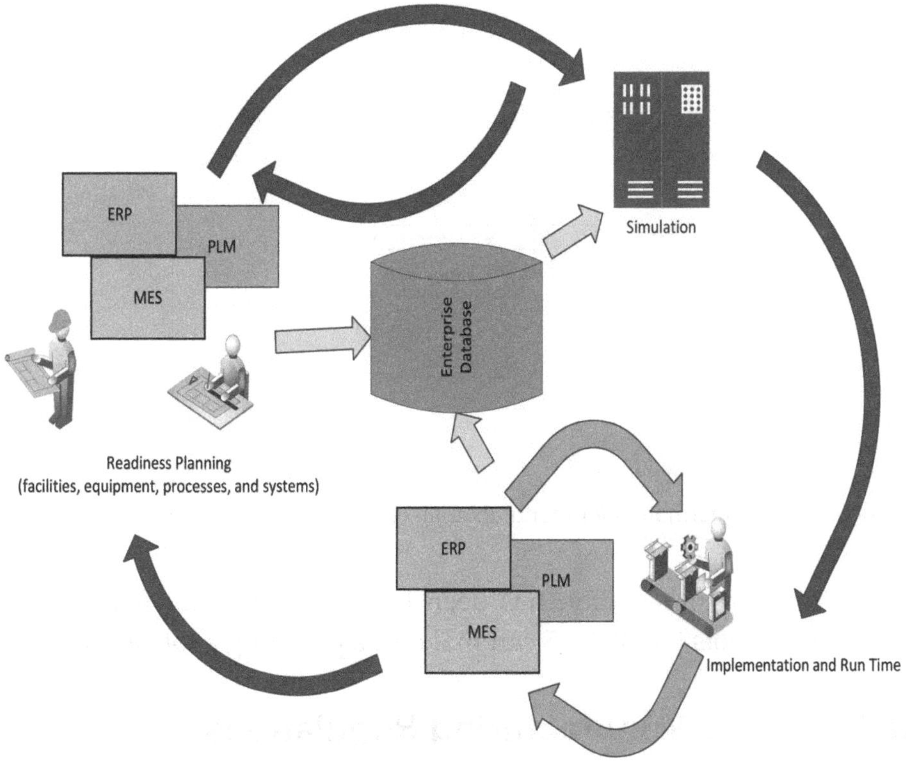

Figure 1-6. Planning, simulation, implementation cycle.

When planning is taken down to the enterprise system level, a lot of information movement must be planned. This planning will require ensuring that information is made available by specific systems, that the data related to that information is made available within the systems that will be the data vaults, and that this data can be transferred to other systems that need the information. Figure 1-7 shows an example of the kind of information planning that is required. The PLM system will likely vault the product-specific information such as the bill of material (BOM), bill of process (BOP), and bill of equipment (BOE). This information is shared with ERP for planning and rough scheduling of procurement and production, which will also be needed for financial planning. PLM will also send this information to the MES to allow for the planning of production execution. Production and inventory data from ERP is also sent to the MES, which will handle the detailed scheduling of material, resources, and production orders, as well as quality monitoring and control.

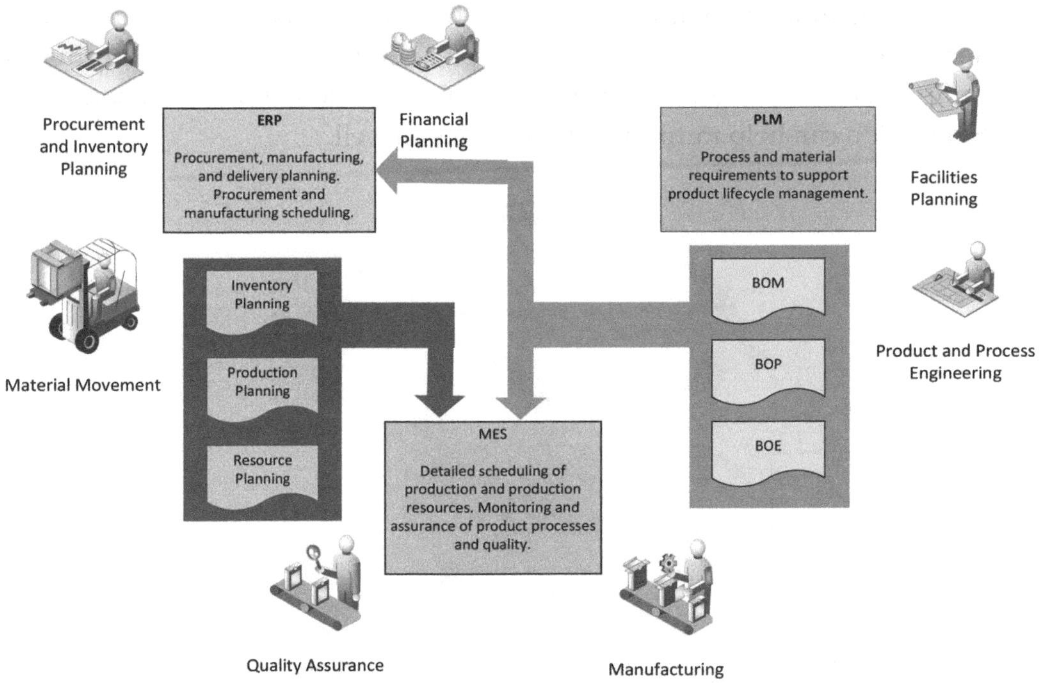

Figure 1-7. Movement of information for manufacturing readiness.

Ensuring that all this information has been taken into account is a daunting task. Using management tools like ISA-95 can make this planning a lot easier.

1.6 The Impact of Manufacturing Regulations

Legal regulations can affect not only how a product is made but also how and what data is collected and organized into meaningful reports. The consequence of not obeying these legal regulations can run into thousands, if not millions, of dollars. Examples of these regulation requirements include the following:

- The ability to trace vehicular air bag and seat belt manufacturing dates and processes
- The traceability of certain chemical mixes in pharmaceutical and food and beverage manufacturing
- Reports describing the process and frequency of cleaning and sterilizing equipment used in pharmaceutical and food and beverage manufacturing
- The traceability of every part and process in aerospace manufacturing

These regulations may be stated at the national or regional level in the country of residence for the manufacturing plant as well as the country to which the product is being shipped. The impact of regulations on the processes within a company must be clearly understood long before the first product is shipped out of the plant.

2

A Functional Overview of MES

To understand manufacturing execution systems (MES), the first question to ask is, "What is an MES?" The answer is best understood in context with the ISA-95 Series of Standards. Figure 2-1 presents an activity diagram of all the activities needed within a manufacturing company. The diagram also highlights the activities that are directly associated with manufacturing operations management (MOM) and includes the operations management of production, quality, inventory (materials and finished goods), and production maintenance. These highlighted activities are primarily managed at the plant level within a company. Activities that are outside the highlighted area are considered enterprise-level activities that are common across all plants.

Recall the ISA-95 hierarchy model from Figure 1-5 that shows the enterprise activities at Level 4 and MOM as primarily the activities within Level 3. As it turns out, MES applications have been designed to support most of the activities within Level 3. This concept is so embedded in industry that Level 3 is frequently referred to as the *MES layer.* Supporting this layer requires systems to maintain a record of the detailed workflow of the production process for each product being manufactured, all major equipment and/or workstations on the production floor, and all operators engaged in production activity. Additionally, Level 3 systems must record the detailed activities performed by production for each production unit during manufacturing.

As we discuss MES, it is important to note that a full discussion on ISA-95 is outside the scope of this book. Many good books have already been published on the standard. References to the standard in this book only serve to provide context to the activity

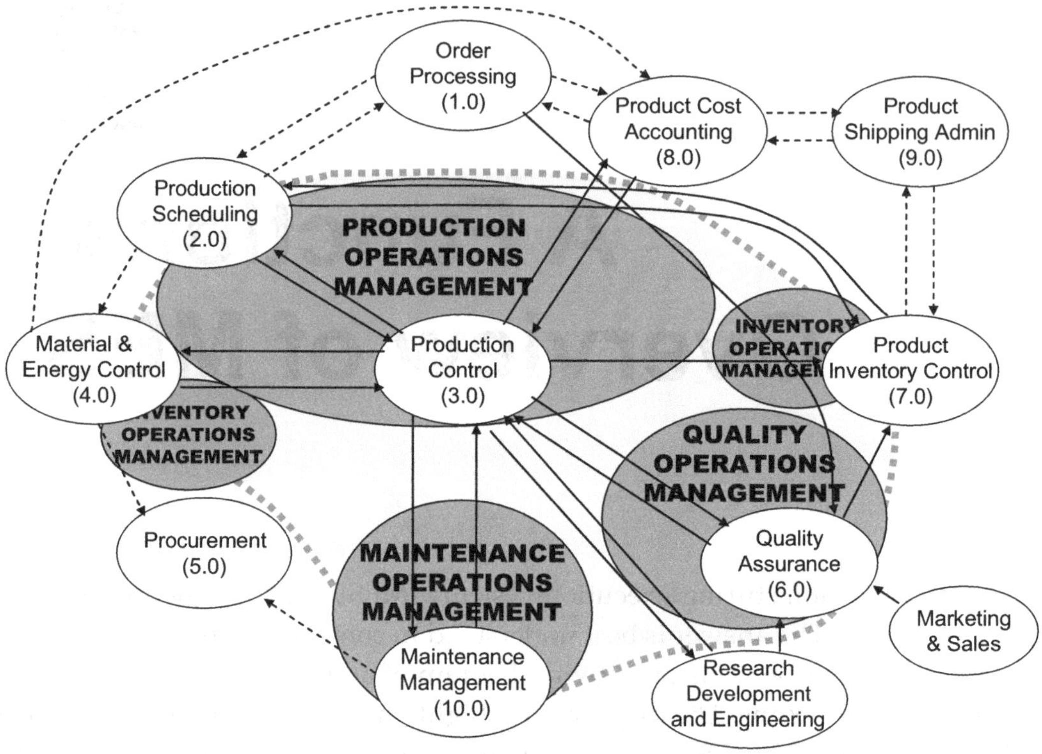

Figure 2-1. Manufacturing activities highlighting manufacturing operations management (MOM). *Source:* ANSI/ISA-95.00.01-2010 (IEC 62264-1 Mod), *Enterprise-Control System Integration – Part 1: Models and Terminology*, Figure 8, Reprinted with permission from the International Society of Automation.

and data models that are used within the MES scope of functionality. For additional information on ISA-95, please see https://www.isa.org/isa95.

The ISA-95 Series explores the functional activities that should happen at each level in the hierarchy. It also defines the structure of data that should be used at each level and exchanged between levels.

Figure 2-2 shows the types of transactions that take place at Level 4 and the resulting data that is part of the integration between Levels 3 and 4. In the ISA-95 hierarchy model, the data and transactions at each level do not necessarily come from enterprise resource planning (ERP) and MES; however, they are the typical applications supporting these transactions. The ISA-95 standards also prescribe a data model for each of the integration branches between Levels 3 and 4. In practice, these data models vary from company to company as a result of the differences among industries, products, and business practices.

Similarly, Figure 2-3 shows the types of transactions that take place at Level 3 with the resulting integration between Levels 2 and 3. Parts 4 through 6 of the ISA-95 Series

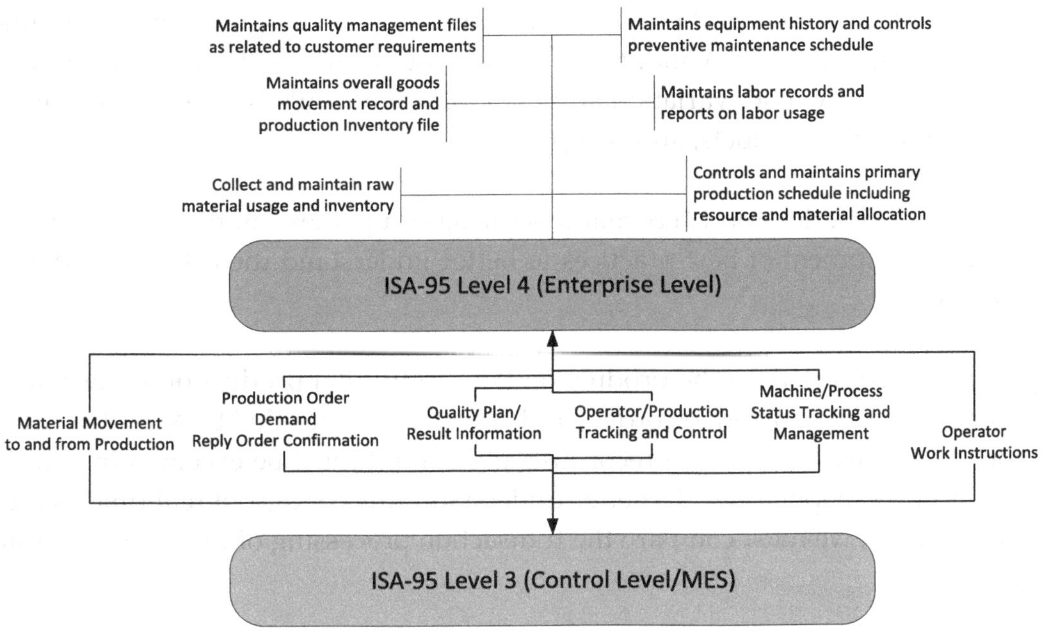

Figure 2-2. Business transactions and integration from Level 4 to Level 3.

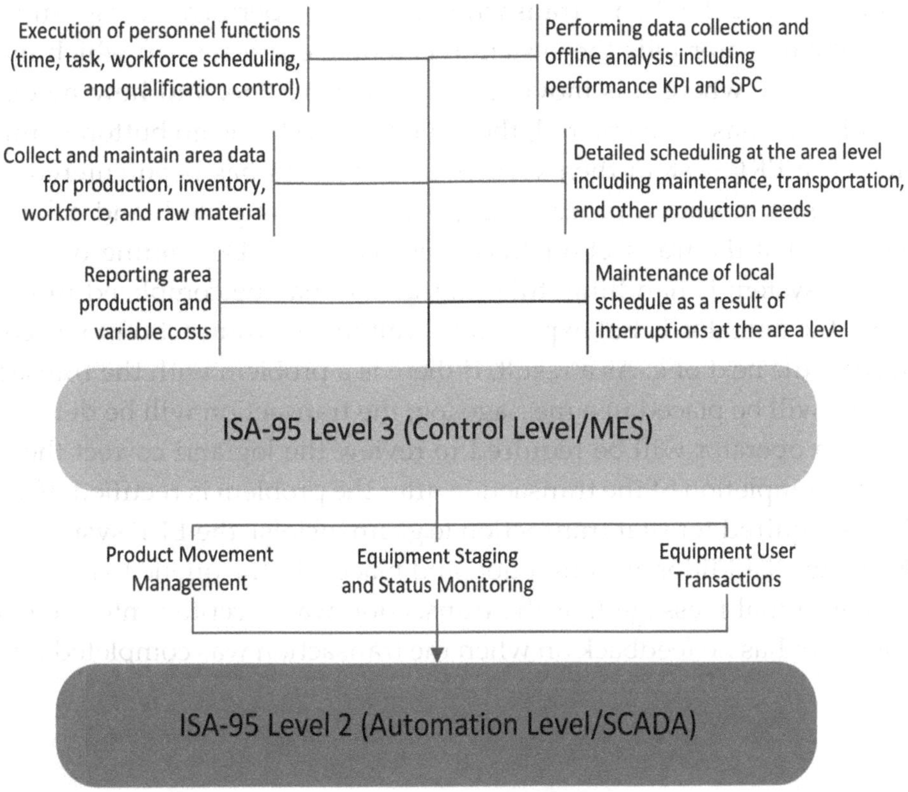

Figure 2-3. Business transactions and integration from Level 3 to Level 2.

describe the data models and provide a full structure for designing the communication sequencing (messaging) and, therefore, the interfacing at these levels. However, there is considerably more variation at these levels as a result of differences in manufacturing processes, products, and equipment.

Using the ISA-95 Series and MES, manufacturing companies can build on the knowledge and development of best practices to better understand the activity on the production floor.

Any activity performed on the production floor to fulfill a production order must be configured to be recognized, recorded, and potentially controlled by MES. For MES to support these activities, all data recording and control must be executed in real time (as the activity is happening). To better understand this concept of real time as it pertains to the MES, we must compare the transaction processing of the MES to a system like ERP.

2.1 Differences between an MES and Other Enterprise Systems

As shown in Figure 2-4, when a transaction for ERP is performed (e.g., an inventory movement), the operator tells the system (via a data entry screen) which material is going to be moved, where it is moving from and where to, and how much is moving. Once this data has been entered, the operator selects the go button to initiate the transaction. The ERP system then verifies that the data has been entered correctly, places the *transaction* onto a processing queue to be completed, and acknowledges to the operator that the transaction "has been accepted." Depending on the volume of activity the system is handling, this transaction may be completed in seconds or minutes, but the operator is not expected to wait for the transaction to be completed before starting the next one. As a result, if there is a problem with the transaction, an error message will be placed in a message log; the transaction will be delayed; and at a later time, an operator will be required to review the log and correct the problem, leading to the completion of the transaction after the problem is rectified. If additional processing is required for that transaction (e.g., financials), the ERP system will automatically create an additional transaction and place it in the queue to repeat the flow. Other than an initial message that the transaction was accepted into the queue, the original operator has no feedback on when the transaction was completed unless they verify it later.

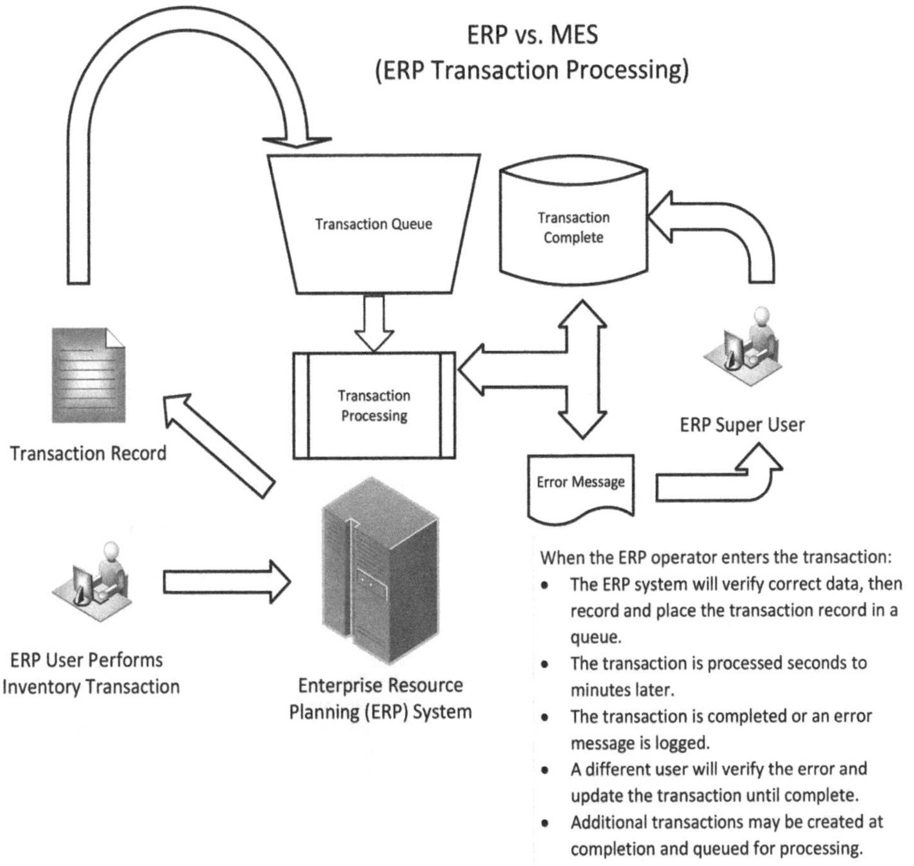

Figure 2-4. ERP transaction processing.

In Figure 2-5, the transaction is not queued, and the operator receives immediate feedback on the success or failure of the transaction.

The expectation from the MES is that the transaction must be validated as correct (the right product, operator, operation, etc.) before the transaction is completed. Additionally, the operator will not be able to perform the next step in the process unless the current transaction is complete.

As part of each transaction, the MES will record the operator, the workstation used, and the operation or equipment that performed the transaction. Additionally, any parametric or assembly data that is pertinent to the transaction will be recorded at the same time. If any of the data is found to be missing or out of spec, the entire transaction will be stopped and *rolled back*. Then, the operator will be notified of the details of the problem and expected to either cancel the step or fix the problem and repeat the step.

With all the data an MES collects and verifies during each transaction, it becomes a valuable tool for Manufacturing and Quality Engineering to use in analyzing the characteristics of any process and/or product from a *good manufacturing practice* perspective.

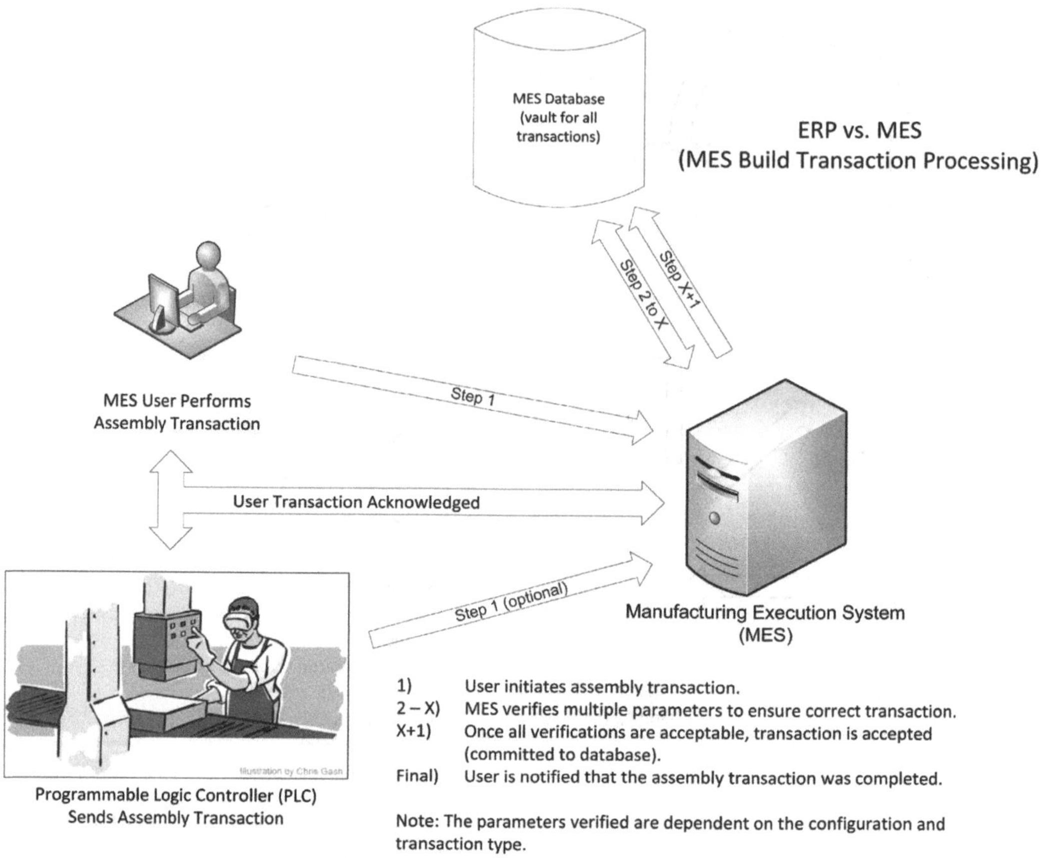

Figure 2-5. MES transaction processing.

Because the MES also verifies every transaction based on the data available at that time, it is also a valuable tool for production. The MES provides immediate feedback on events that may affect the quality of the product and can be configured to possibly stop not only the particular production unit but also the entire production order if the problem is significant enough.

In some production environments, waiting a few minutes for an ERP system to respond in the same manner (if it can respond this way at all) could mean the loss of thousands of dollars.

2.2 General Outline of MES Functionality (Grouping the MES Functions)

The functionality of an MES can be categorized into three main blocks (see Figure 2-6) that are based on a hierarchy or the maturity of the manufacturing processes. The blocks include the basic track-and-trace capabilities, the facility management functions

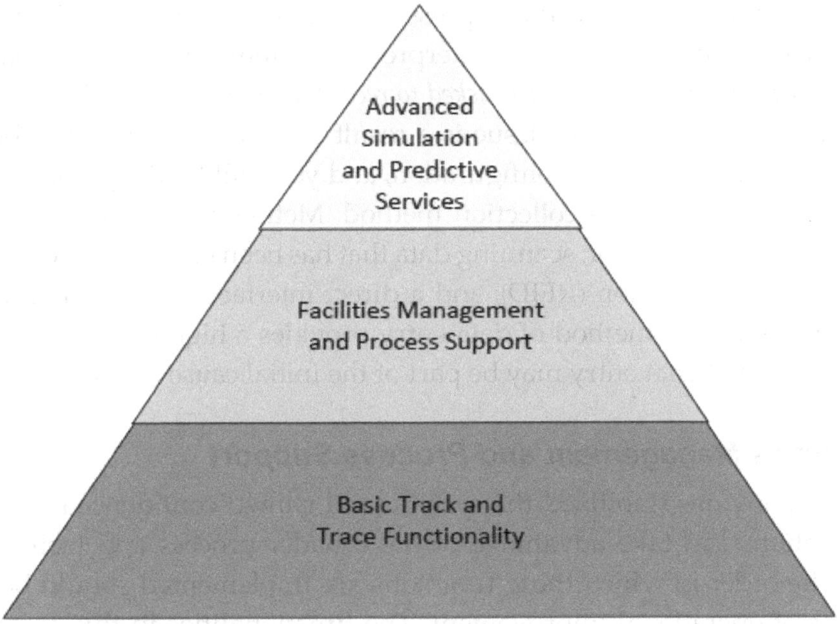

Figure 2-6. Three categories of MES functional support.

found in companies with more mature process management, and the advanced functions used for simulation and predictive analytics.

2.2.1 Basic Track and Trace

The primary functionality found in almost all MES is the capability to monitor, control, and report on the movement of a product through the production process (and with that, the status of any production orders). This includes tracking the genealogy of any given production unit as it is being manufactured. The purpose of this basic functionality is to ensure that each production unit (which can be represented by a batch ID or serial number) has completed the correct process, has been manufactured with the correct material, and has been processed by the manufacturing floor at the correct time.

Companies that are new to MES should start by implementing the basic track-and-trace capabilities because they are the easiest and least costly to implement. They also provide a quick return on investment with stable and repeatable production processes and easily accessible data that can be used to launch continuous improvement initiatives. These capabilities can be implemented with little to no integration with equipment, and they create a system that is tolerant of process changes. Implementing this level of a system will create visibility of where processes are deviating and enable the company to stabilize both the processes that the system supports and the use of the system. This enables the operator to gain confidence in what the system is telling them. The greatest block to system acceptance at this level is the impression that the system is not giving a true picture of production. The correct way to interpret the system at this level is that *the system is providing a true picture of what it was asked to record.* If there is a problem with the data the system is providing, then the issue is a result of how the system has been configured to record that data. Fix the configuration, and you will fix the problem data. Some issues are a result of the data collection method. Methods implemented will include manual entry from an operator; scanning data that has been barcoded, including 2D and radio-frequency identification (RFID); and a direct interface to the equipment or tools providing the data. Each method of data entry provides a higher level of accuracy and repeatability. Manual data entry may be part of the initial cause of bad data.

2.2.2 Facilities Management and Process Support

After the company has stabilized the system (and gained confidence in the data provided), operations can take advantage of the broader process and facilities support functions. The order in which these functions are implemented should be driven by the priorities of operational improvement. The functionalities in this category might include equipment status monitoring, real-time operations reporting with Andon screens, and process event triggering. These functions can provide valuable insight into the *current status* of production orders or processes as changes occur, enabling operations to respond early when a process or order execution deviates from what is expected. The company will want to implement more detailed data collection regarding the process and equipment input. As a result, this level of MES functionality will require a great deal of integration with the equipment on the floor. This integration can be implemented quickly and extended if Manufacturing Engineering also implements standardized communications with the equipment-level applications. Within this level of MES use, companies can also consider implementing operational measurement methods such as overall equipment effectiveness (OEE) and statistical process control (SPC), as well as other quality management methods. All detailed scheduling and tracking production line stocking activities are also part of this level. These activities are covered in later chapters.

2.2.3 Advanced Simulation and Predictive Services

Once a company has been using data via equipment integration and has a good understanding of the equipment and process characterization, it is ready to start implementing predictive analytics and process simulation. Trying to implement these functions too early can lead to support staff being overwhelmed by false issues without having the data available to clarify and eliminate the problem. As a result, care should be taken to implement this final category only after the proper characterization data is available through equipment integration.

Once Operations has interfaced the MES to equipment on the production floor and a history of data has been established, it becomes possible to use this data along with statistical analysis methods to recognize and predict problems with equipment performance, thereby preventing product quality issues as well as unscheduled downtime due to equipment failure. As part of the MES data collection capability, transactions such as component assembly, material fabrication, and functional testing can be correlated to sensor data from the equipment by product, user, or processes. This will give Operations the ability to predict when failure is likely to happen based on the number of operational cycles any piece of equipment has performed. This information can also be used to model more effective preventive maintenance programs, providing for longer mean time between failures (MTBF). Making a detailed set of sensor data from equipment available is a major driver in the latest Industrial Internet of Things (IIoT) concepts that the industry is investigating.

Part of the production planning process is to try to estimate the capacity of a new facility (or a change to an existing facility) before investing in the actual change projects. For this kind of estimation, companies will frequently turn to simulating the manufacturing process. For products that share a production line with similar products, simulation can also be used to analyze production volumes to determine at what volume a product should be moved to a dedicated line. To gain valuable insight from these simulations, Operations must have considerable knowledge about the equipment performance and the operational steps that will be used in the new facility (or line). An MES that was previously in place can be an excellent source of data to be used to complete the needed characterizations that make simulation valuable.

2.3 General Support of Operations

In addition to the categories previously listed, MES can be implemented in systems that support discrete manufacturing and process flow manufacturing. All MES will provide support for these industry segments in varying degrees in a manner that is typically geared toward specific industries like the pharmaceutical or automotive industry. Note that designing an MES that can support all industries equally is simply not practical from a system development perspective. As a result, Operations must be keenly aware of the manufacturing processes and requirements within their industries when looking at MES capabilities. In addition, configuring the MES requires a significant understanding of the data that will be stored in the database as well as the context of how that data will be used from an operations perspective. If this level of understanding is not available at the time of implementation, there is a strong possibility that the MES configuration will misrepresent the production floor and provide misleading information to Operations or limit the system's use as the company matures and grows.

It takes time and considerable experience with MES to design a system that supports the broadest range of functions and supports them well.

Another way of looking at MES functionality would be to group the functions by the business processes they support. The actual functions that are implemented for the processes may vary depending on the system. However, the way the functions are implemented can have a dramatic effect on the level of support that they provide as part of the full scope of the MES solution (which will be discussed in later chapters).

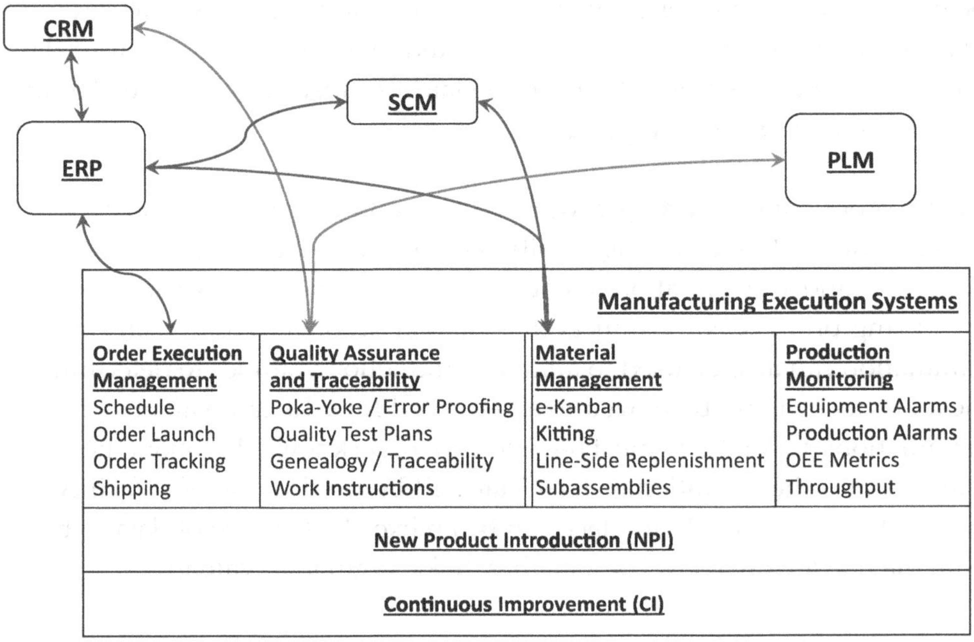

Figure 2-7. MES categorization by business function.

In Figure 2-7, the MES functions are displayed under the business processes that they mainly support or influence, and the enterprise systems related to each specific business process. Of course, some functions can support multiple business processes, but this is not shown for simplicity. The MES functions include the following:

- **Order execution management** – This group of functions includes scheduling material delivery to production and product manufacturing, tracking the quantity and rate of order fulfillment, and tracking the product from both a material consumption and a process throughput perspective. In addition, as the MES tracks the movement of material and work-in-process (WIP) inventory, the data collected by the system includes the time, user, and operation/resource (and other data elements) in the transaction record that was recorded for that movement. This data is used to understand cycle times and *planned versus actual* reporting, which is used by ERP for improving the planning process. It includes the device history reporting that is used for continuous improvement, warranty management, and content references for shipping containers.

- **Quality assurance and traceability** – By tracking the material genealogy (the material batch ID or serial number) that was used during production and recording the process pass/fail results (test or inspection), the MES can provide insight into the performance of material suppliers. In addition, an MES can capture WIP inventory nonconformance data for use in correlating types of failures with components or production steps. This data can be used for continuous improvement and to ensure that a nonconforming WIP is either reworked/ repaired or properly scrapped so that it cannot be sold. MES functions that support quality assurance cover a broad range of capabilities and requirements, including the implementation of poka-yoke principles and work instructions. These are covered in later chapters.
- **Material management** – These functions manage the movement of material from the warehouse to the production floor from both a quantity and a genealogy perspective. Functionality from an MES perspective may include support for *min/max* for line-side stocking, e-kanban, and kitting. Other inventory management methods may also need to be supported. These requirements are covered in more detail in later chapters, primarily in Chapter 4.
- **Production monitoring** – These functions monitor the rate and efficiency of production processes and signal to the production staff when either equipment or process events occur that may affect the production schedule. This business process may be seen as a part of order management, but it has particular requirements that are worth a more detailed discussion. Access to the *should-be* configuration of a production floor also enables the MES to compare it to the current real-time status and prevent transactions (production steps) from being performed by the wrong equipment or the wrong operator.
- **New product introduction (NPI)** – All four business processes previously described are required to help launch the manufacturing of a new product or upgrade an existing product. The NPI process is used to analyze and verify component/material performance characteristics, validate that a product is going to perform as expected through production, and ensure that the product can be manufactured at an acceptable rate of production. Each of the previous business processes must be in close coordination with the others and may include updating the MES configuration to ensure the proper launching of a new or updated manufacturing process, new equipment, and/or a new or updated product.

Notice that MES functions have links with both ERP and product lifecycle management (PLM) and can extend to other enterprise applications. While ERP schedules the various aspects of the manufacturing enterprise, the MES is responsible for monitoring and managing the actual execution of the schedule. Also, the manufacturing process requires input from the PLM on the design and engineering intents for the product to ensure it is manufactured properly. The input can include key process parameters (e.g., temperature, pressure, and manipulation cycles in some processes) and the work instructions required by the MES for execution. During the manufacturing process, data is generated that can be sent back to the PLM for product support information and, later, to support product improvement activities. This is how ERP, the PLM, and the MES are all interconnected in the manufacturing operations management concept presented in Chapter 1, "Introduction, Overview, and the Current State of Manufacturing."

The general function of the MES, regardless of the industry, is to collect data specific to each step in the manufacturing process and then use it with other data (or instructions) to guide, monitor, and sometimes control that same manufacturing process in real time. The MES also reacts to situations that may affect product quality faster than is humanly possible in a manually managed production floor. The level of awareness that an MES has of a manufacturing process is highly dependent on what the system is configured to monitor and on the amount of control it is configured to provide. Whether that control is determined to be automated or is implemented to alert and guide an operator also depends on how the system is configured. Care should be taken when configuring the production floor within the MES because the configuration determines what data the MES can provide. Poorly configured production environments can make it difficult to mine data for information and limit the confidence in and utilization of the system.

The question that now comes into play is, "when should the system control a situation, and when should the system only provide guidance?" Simply put, the answer is to *allow the system to control the situations that make sense.* The remaining chapters of this book will provide guidance for Operations on how to answer these questions for their specific situations.

3 Order Execution Management: Details and Benefits

3.1 Introduction of Order Planning and Execution

A major activity for managing manufacturing companies is determining which products to make, how many, and when they are going to be available. This process starts at the executive level with the rough-cut planning necessary for sales and operations planning (S&OP) and is performed a year or more in advance. Then, for the most part, the process ends with scheduling the correct sequence of products to make the most effective use of resources on the production floor and is determined at the level of products per shift. This planning function is further complicated due to recent trends of mass customization and the demand for shorter delivery times.

Consider the automotive industry as an example. As part of an order planning process called *in-sequence*, suppliers of parts to the original equipment manufacturer (OEM) manufacturing lines are expected to have the capability to receive a schedule for a product of a specific configuration (color, optional features, etc.) within a week of the expected delivery, deliver a set of semi-assembled parts to an exact location on the OEM production line within a half-hour window, and deliver the set in the correct sequence required by the OEM. The supplier must schedule the receipt of materials and resources to the production floor and load the delivery truck in enough time for it to drive to the OEM location and deliver the material within the prescribed window regardless of local traffic and weather conditions. In many circumstances, the end customer can change their mind about certain options as close to 24 hours before delivery to the OEM assembly line. If the supplier gets the configuration or sequence wrong or if there is a nonconformance on delivery, the supplier then faces significant fines from the OEM to correct the problem.

Most manufacturing sectors are not as tightly scheduled as the automotive industry, but all industries have their own constraints and problems. Planning and scheduling for a manufacturing floor can be a stressful activity, considering the scheduling, logistics, unexpected events due to materials not conforming to expectations, and equipment performance deviation or failure that can occur. Planning departments struggle with keeping production flowing, and information that identifies a potential issue is best provided to the department as soon as possible. In many cases, seconds can make the difference between a successful shipment and a major failure. The recent push by Industry 4.0 and Smart Manufacturing initiatives is an effort to (1) keep unexpected events from happening and (2) provide notice of events that happen (and the effects) to the entire length of the supply chain. In this chapter, we will discuss the data and information needed to keep things moving on time and how a manufacturing execution system (MES) can help achieve this goal.

3.2 Order Execution before MES

To understand the benefits of having an MES in order management, take a step back and look at order management issues before the advent of MES. Many companies still work in this environment, and if a company has well-defined business-level processes and relatively simple product order management, this scenario may be acceptable. In this section, we will review the planning process at a high level and provide input on the issues that MES helps to solve.

Figure 3-1 illustrates the planning process for order management:

1. Determine the current capacity of the manufacturing environment and compare it to the marketing plan as the initial S&OP, which is needed to determine the master production schedule (MPS). Part of the concern of S&OP is a result of the *hidden factory*, where up to 30% of the capacity is consumed by repair, rework, and expediting processes that are not tracked within enterprise resource planning (ERP).

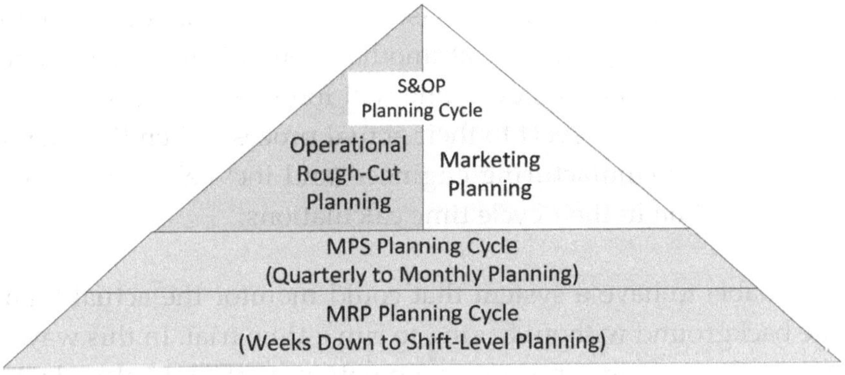

Figure 3-1. Enterprise planning layers.

2. From the sales estimate, determine the MPS; in some cases, the MPS may be partially replaced by actual customer orders.
3. Run material resource planning (MRP) to determine (1) what material is needed and when to fulfill the plan and (2) when the order must be released to Production to deliver (or have finished goods in inventory) at the right time.
4. When the timing is right, release the purchase orders for materials and/or production orders for manufacturing to initiate the delivery of finished goods to the warehouse (or the retail outlet) by the expected delivery date.

The timing of the order release is based on (1) the lead time to receive the material and manufacture the product and (2) when the product is expected to be in finished goods. With this information, the planner can simply select the "Release" button on the ERP screen at the right time. The material will arrive right on time, the manufacturing floor will make the product when needed, and the finished goods will end up on the shelf right when they are needed—right? (*Yeah, right, this is the way it happens all the time.*)

3.2.1 Determining the Lead Time

Before MES, manufacturing engineers had to perform a trial order (or periodic sample order) to calculate and verify the lead time of a product. The problem with this method is that each product has its own lead time that is subject to change whenever there is a change in the product or the process for the product. Each time there is a change, the manufacturing engineer should requalify the new lead time. However, this procedure is time-consuming, and as a result, it is typically not performed enough to provide meaningful data for the planning group.

An additional problem that occurs is a result of the difference between the published process and the actual process. This frequently happens when the production operators have implemented *shortcuts* to the process (sometimes making it easier at that point in the process but unknowingly harder at another point). When manufacturing engineers are doing a time trial order, everyone will follow the process because they are being watched, but they will revert to their actual process when the trial order is finished. As a result, the manufacturing engineer (and inevitably the planning group) uses skewed information in their cycle time calculations.

It would be valuable to have a system that could monitor the actual manufacturing process in the background without having to run a time trial. In this way, data would be provided regarding product/process combinations that could also identify changes in the processes and why they have occurred.

There are several other issues related to processing capability beyond lead time. Many of these issues will be covered in more detail in other chapters of the book.

3.2.2 Other Issues that Arise

The main rule about production plans and schedules is that they are only good until they are published. It is not a question of "if things will change;" it is a question of "when things will change, and by how much." Many problems can arise with fulfilling a production schedule; however, for the sake of simplicity, we can narrow them down to three main contributors:

1. **Material shortage** – Material that was planned to be available but was unexpectedly not available when needed for various reasons

2. **Equipment availability** – Equipment (also referenced as a *resource*) that was scheduled to be used during a production run but was unexpectedly not available as a result of downtime

3. **Production operator's availability** – Skills required to perform critical functions that were either unexpectedly not available or redirected

The key term in each contributor is *unexpected*. In the real world, things happen, and considerable effort goes into trying to deal with unexpected events. In addition, despite companies' attempts to prioritize communication with the planning group, inevitably these events are not immediately communicated, and products fail to be completed on time, resulting in an unhappy customer. Therefore, there is not only an unexpected event that disrupts the production schedule but also a failure to communicate it in time to recover from the problem (if recovery is possible).

From the planning perspective, it is essential to first identify when an event occurs that will affect the expected planning schedule; then, once the event is known, it is necessary to determine the extent of the problem. Getting this information quickly is important to the planning group (as well as other groups) because it is needed to determine how much the scheduled delivery is going to change and what can be rescheduled to minimize the effects of the problem. To compound the problem, getting this information is difficult because these questions are frequently in competition with other questions that the rest of Operations is trying to resolve for the same problem.

It would then be valuable to have a system that could monitor the progress of a production order, detect when there is an issue with its scheduled completion, and provide the information required by the planning group while simultaneously providing information to other groups within Operations—all in real time as events are happening.

3.3 Tracking Order Execution with MES

At any time during production, the Manufacturing Engineering department is looking for details on the process capability, the Quality department is looking for the statistical status of quality parameters for all products, and the planning group is looking for insight into the current status of all production orders. An early warning when things are going wrong is of value to all of them. From a data perspective, these functions mostly rely on the same data elements that are only reported in a different context. With very little *tweaking* of the data collection steps, these functions can be supported with the same base data structures from an MES database.

3.3.1 Basic Track and Trace

The primary function of MES has always been to track products during the production process, as illustrated in Figure 3-2. In doing so, manufacturers are able to determine the status of a production order at any time during production. The fact that an MES identifies and records all production-level transactions (e.g., starting an operation for a serial or batch number, assembling each component, and/or monitoring each fabrication step and the operation completion time) and does so in real time, enables the MES to have a unique view of the production environment. If the database of an MES has been implemented properly, the MES knows when each production unit went through each operation, which operator was logged in at that operation (presumably the operator doing the work), and at which workstation the work was performed (as long as the operation name and the workstation are identified differently). All this data is then correlated to the time that each action was completed, and the MES can frequently collect this data in the background while the work is performed. Although it is covered

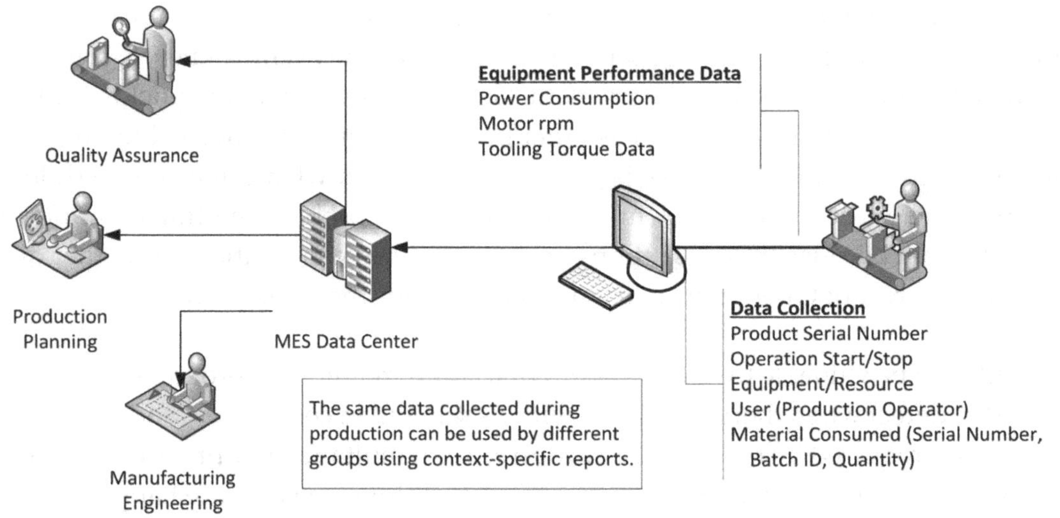

Figure 3-2. Track-and-trace data collection.

in greater detail in later chapters, the data that is collected throughout the entire production process is also available for comparison to performance expectations. Most of this comparison will be determined by the master data that is set up by the various functions of manufacturing within the MES.

3.3.2 Getting the Early Warning

Using an MES to manage the manufacturing process (as defined by the product routing) and the manufacturing bill of material (mBOM—a detailed list of the material and quantities along with the exact operation and step within the operation that the material is to be consumed), the MES knows the steps to be completed and the materials to be used during the production process. However, as part of the product routing, the MES can also have the standard cycle time for each operation and the wait/move time between operations. It can therefore calculate the target time for completing the order. From the detailed records of where products are within the process, the system can determine the current status of the order and if production is currently on time, leading, or lagging behind the estimated time to completion. By using free-running dashboards, the production and planning groups can have a detailed understanding of each order in comparison to its expected completion and, with the static cycle times that are on record, the current expected completion time. For some companies, there is an urge to use the average cycle time over a specified period of time in estimations for order completion instead of a static standard time. The problem with using the cycle time average is that it can change significantly as a result of short, one-time events that should not be considered as part of the planning process, therefore creating erratic feedback in the planning process.

3.3.3 Maintaining the Cycle Time Reference

As discussed throughout this section, considerable effort goes into planning when to release production orders, and this activity is highly dependent on manufacturing cycle times. However, there are many reasons why the cycle time should change over time. Process improvements, product changes, and even familiarization with the process can have a significant impact on the actual cycle time for a product/process combination. If the MES is implemented properly, the system will have a record of the actual cycle times, and that data should be made available to the MRP system. By using the MES data to create the standard cycle times and properly manage the process of updating the cycle time record, Manufacturing Engineering can easily provide and update the information that ERP uses during MRP. Care must be taken as to when and how often this data should be updated to ensure that the MRP runs do not become too erratic for effective use. As an additional benefit, this information can then be fed back into the rough-cut capacity planning to improve the initial S&OP.

3.3.4 Continuous Monitoring of Production

Dashboards are essentially static reports that are run in continuous loops over a predefined period. When the dashboard (or another report) determines that an order is lagging by more than the specified amount, the system can be programmed to notify people by email (or any other form of communication) that there is a problem that requires their attention. Using reports that have already been developed (all accessing the same data in one context or another), each operational function (like the planning group) can focus on the particular information that will determine their course of action. With a skilled MES implementation team, these reports can be defined to ensure that each group is getting different interpretations of the same message.

The context for the other functions will be explored in more detail in the following chapters. However, for the purpose of monitoring the production order status, the dashboard provides a detailed view of the status of all orders in production, the capability to continuously estimate the time to completion, and the ability to provide alerts to required groups of people the moment an event occurs. These capabilities are beneficial to the planning group when it comes to managing the on-time delivery of products through the production process. These alerts can come in the form of emails or messages or may be part of a dashboard that graphically displays the planned and current estimate of an order. As dashboards become more popular, this is becoming the more common form of an alert.

3.3.5 Labor Tracking

An additional part of the planning process is to estimate the direct labor used during the production of a part. Again, the efforts of Operations are typically collected by

Manufacturing Engineering during trial production runs. In the same manner that the MES provides visibility of material consumption and order status, the MES will record the amount of time that an operator at a specific operation has been working on a given serial or batch number. As a result, the MES can provide the same detailed data on direct labor that it provides for order status. In this scenario, care must be taken in developing the user interface and production rules setting that will prevent this kind of data from being skewed. Having relatively accurate information on direct labor costs is a huge benefit to operations cost control. These requirements should also be a factor that is controlled by the implementation team at first and as part of the ongoing maintenance of the MES configuration.

3.3.6 Control and Release of Suborders (Handling Subassemblies)

When using an MES in a production environment with subassemblies that are manufactured in the same facility, there are options for managing these orders. From an ERP perspective, there are two general ways of handling subassemblies:

1. Handle the subassembly orders as independent orders, in which case there is no difference to the MES. Regardless of the subassembly orders being released at the same time, reporting and managing the orders are independent of each other. This is usually the management method when a subassembly is used for multiple products or if it has a significantly different rework or scrap rate than the parent product.
2. Handle the materials of the subassembly as line items for the parent order, with the subassembly being manufactured as part of the process for the product. This is used when the subassembly is used exclusively with a parent product (referred to as *phantom assemblies*) or if the subassembly must be directly matched to a specific serial number of the parent (referred to as *pegging*).

In the first option, there is no difference in the way MES handles orders at the subassembly level and the parent level. The order for each BOM level of the product is released to the MES, and each is tracked and reported independently.

The concern with using phantom material from the ERP level is that at the production floor level, Operations loses the ability to uniquely identify the subassembly and therefore loses the visibility of tracking work on the subassembly itself. If the phantom level (the actual subassembly) has a minimal material and labor charge against it, then even the lack of visibility is not a problem. Alternatively, *phantoming* may be a good idea from a planning perspective, but it is not good for the Manufacturing Engineering or Quality departments that want individual tracking and build history. The MES, even at a basic track-and-trace level, can still provide an option to manufacturing operations.

When the parent order is released from ERP (with the phantom material) down to the MES level, the interface to the MES can separate the phantom material into an independent order ID. The MES order ID would not have to be reported back to ERP, but it is still able to provide a unique identifier that enables that part of production to be independently isolated and analyzed at the MES level. The same order release process can also be used for pegging, where there is a need to directly match a subassembly with a specific serial number of the parent. As the MES knows which child order was released from the parent order, the MES can provide the control to ensure the child is exactly matched. The physical process would also require methods for aligning the child and parent build times. Fortunately, there are a limited number of industries that use pegging. However, a type of pegging function is used within the automotive industry for sequence management. For the purposes of the MES level, any sequence ID that is provided from the ERP level can be used as the equivalent of an MES order ID and provide the correct level of control.

3.3.7 Shipping and Advanced Ship Notice

A function that can be very useful in an MES is *container management*, which is used for the preparation of product skids. During the manufacturing process, the MES can be given the additional task of identifying a skid (or a container) of finished goods and using that information to create a shipping manifest or bill of lading (BoL). By giving a container ID to each container that is created and tracking all serial numbers (or batch IDs) that have been placed in the container, the MES has all the information needed to create the BoL, and as part of the report back to ERP, can provide the detailed contents of that container to be included as part of the shipping documentation. If the MES is implemented properly, tracking the finished products that go into a container is no different than tracking the components that go into a finished product.

Just as the MES tracks this data as part of its normal operation, so too does it track the *build* of a container. Whether it comes from ERP or directly from the MES, once the container has been completed and is ready to be loaded onto a truck, either system can use this data to create an advanced ship notice to be sent to the next-level customer. With this data already made available to the customer, the receiving process at the next level can be much more streamlined.

3.4 Additional Concepts

In Chapter 2, "A Functional Overview of MES," the general context of the interaction between the enterprise level (ISA-95 Level 4) and the MES level (Level 3, the control level in ISA-95 terms) is highlighted. An important part of a good MES implementation is to respect the demarcation of activities that should be performed by the different levels. As it pertains to order execution activities, the responsibility of an order release (i.e., when the order should be scheduled) belongs at the ERP (enterprise) level and should be maintained at that level.

Once an order has been released from ERP, the detailed scheduling of a production order (when to start producing product to an order) should be managed as a part of the real-time status of the orders in the MES. As a result, many MES provide a module to specifically handle the detailed scheduling function. The detailed scheduling function at the shift level should not be used to override the order release from the enterprise level.

When a manufacturing company has gained considerable maturity in the process characterization of order execution, it can be beneficial to automate the production initiation phase of an order. It is then valuable to have the MES automate material replenishment, which is covered in Chapter 5, "Quality and Traceability: Details and Benefits," but the function of initiating the start of production must be monitored closely and should not be automated quickly.

3.5 Order Execution Benefits

The benefits covered in this chapter include the following:

- Detailed tracking and information on the status of production orders in real time
- Early warnings regarding schedule deviation as part of the status reporting function
- Better management of cycle time planning without having to run special test orders
- The ability to use real-time dashboards to report order status
- Detailed tracking of labor usage to improve financial planning efforts
- Multiple options for order release of subassemblies

The benefits of MES reach beyond monitoring order status and helping improve planning data for MRP and MPS estimations. In the following chapters, we will examine other business processes to review additional ways MES is of value to manufacturing companies across the entire enterprise.

4 Material Management: Details and Benefits

Another key component in manufacturing is *materials management*. This refers to the raw materials and parts used to make the end product that must be ordered, delivered, and managed by the plant. Management includes not only ensuring that the parts are delivered to the proper production line station at the right time and in the right order so that production is not affected but also managing the part or material inventory. Without this process, ensuring that the right material is at the right place at the right time in order to fulfill the production schedule becomes a difficult task and will lead to production delays or surplus parts inventory (and the excessive cost of carrying inventory).

4.1 Material Management Business Process

4.1.1 Overall Process

Material management is executed through supply chain management (SCM) processes. This means that before the material or parts can be managed within the plant environment, certain business transactions must occur between the manufacturer and its parts and material suppliers. Figure 4-1 shows an overview of the SCM processes that occur before the material arrives at the plant location. Depending on the industry,

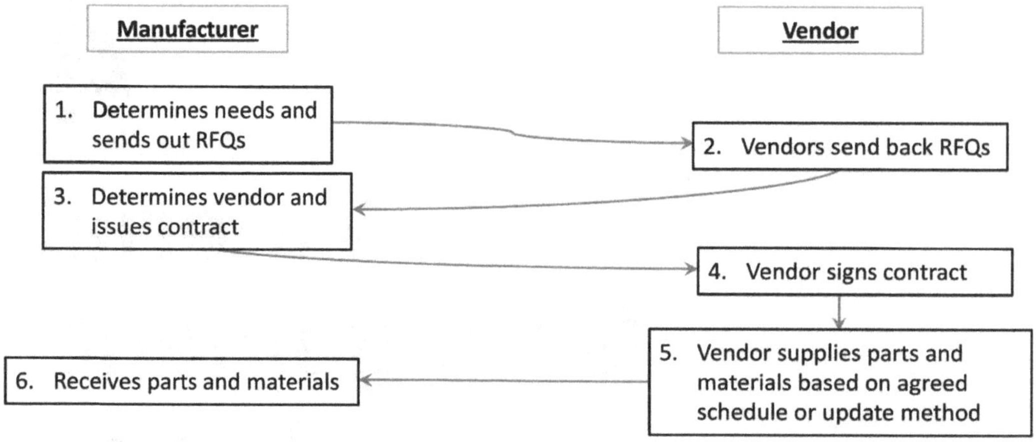

Figure 4-1. Supply contract process.

these processes may vary; however, a discussion of such differences is beyond the scope of this book.

4.1.2 Material Management before MES

To understand the benefits of having a manufacturing execution system (MES) in material management, we will take a step back and look at an example (in discrete manufacturing) of the issues in material management before MES was used. It should not be surprising that before MES, all material management was conducted manually and usually monitored at the enterprise resource planning (ERP) level in aggregate quantities. To follow the lifecycle of a subpart or raw material in the plant, every status check in the process must be manually updated.

Starting with the plant receiving the part or material, the inventory must be manually updated, typically on a hard copy checklist. When the part or material is physically transferred (or moved) to the bulk storage or warehouse area, the inventory must be manually updated in ERP. Parts or materials typically are not transferred to the part pick or staging area until the inventory there drops below a certain threshold and must be replenished to its maximum threshold. This also must be manually monitored, and again the inventory must be manually updated. Operations methodology like kanban has been implemented to try to make this easier. Parts and materials leave the part pick area either when a manual pick list is generated or when a manual replenishment is triggered from the production line workstation. Throughout this process, inventory must be up to date so that when the next order for parts or materials must be sent to the vendor, it can be triggered at the proper time.

All manual triggers mentioned depend on a person being designated to trigger the event at the proper time, which is when the inventory for that specific workstation hits

its threshold. This leaves a significant opportunity for error, not only for replenishment triggers but also for accurate inventory updates. The consequences of improper replenishment triggers or inaccurate inventory levels include the line workstation running out of parts during production, which shuts down the line, or the plant running out of parts, which shuts down much larger aspects of production. In a manufacturing environment where the delivery of the end product is *just-in-time*, these production stoppage events must be minimized as much as possible, or the company may face *stockout* fines from its customers.

Another consideration in this manual material management support model is that certain material management activities, such as part kitting or subassembly management, are more difficult to manage because all data points are tracked and managed (i.e., organized, collated, and reviewed) manually. Depending on the amount of data required, this takes a considerable amount of time and labor to accomplish. In an effort to prevent stockouts, manufacturers will maintain considerable levels of inventory and set the trigger levels high enough to provide for delivery lead time, and/or the manufacturer will implement methodologies such as inventory cycle counting, which also require considerable labor. This costs the company both the money invested in the inventory (which consumes cash flow) and the real estate for storage (which consumes production floor space).

In addition to this discrete manufacturing example, material management in process manufacturing had its own issues prior to the use of MES. In many instances, the raw material used for process manufacturing (pharmaceutical, chemical, etc.) required specific batches of material for a particular production run. Operations had to ensure that the correct batch of material was used for a specific order, and depending on the size of the order, they may have needed multiple batches to complete the order. Batch details, including which batches were used, how much of each batch was used, and batch characteristics, had to be tracked based on process and end-product requirements. To complicate matters even more, batches of material may have had expiration dates that had to be managed. Not managing these batches properly could have resulted in scrapping entire production runs.

4.1.3 ERP and SCM

The business process of material management begins with ERP and SCM. This means that the contracts with the material and part vendors must be in place first. These contracts must include the following information:

- Production schedule
- Quantity required

- Delivery schedule
- How the material consumption is tracked (and if automated reordering is possible)
- Whether the parts are bulk or end-product specific in nature

This information is only a partial list. Details on ERP and/or SCM operations are beyond the scope of this book. All these values must be carefully calculated based on the initial part or material consumption rate to maintain the proper in-plant inventory and to avoid stoppages in production due to a part or material shortage. If the consumption rate for the manufactured part changes during the lifecycle of production (whether it is increased or decreased due to changes in demand), the supply chain must adjust accordingly, from the vendors down to the production line. This will require careful monitoring by the appropriate people at both the vendor and the plant levels. As such, an MES can automate these processes (especially in a high-volume manufacturing environment) and can ensure each of the line-side part bins is replenished in a just-in-time manner. This is done through line-side replenishment processes. These processes operate independently of ERP and SCM, but they provide part consumption updates to both. If timely updates do not occur, the risk that reordered parts or material will not arrive at the plant at the right time increases and a production stoppage could occur. When the line-side stocking (LSS) processes have been stabilized and the inventory counts controlled to an acceptable level, the company can justify lowering the need for inventory counting, saving both time and money.

At the production floor level, material management is used to ensure the proper parts are in the right sequence on the production line to support production. Doing so gives the added benefit of streamlining the whole SCM process, from the vendor to the part or material installation point on the production line. ERP and SCM cannot operate successfully without material management and vice versa.

4.1.4 In Plant

The lifecycle of a part or material in a discrete parts manufacturing plant is shown in Figure 4-2.

The lifecycle of a part is as follows.

1. The plant orders a part (see Section 4.1.5, "Material Status (In Plant)," for more details).
2. The vendor ships the part to the plant.

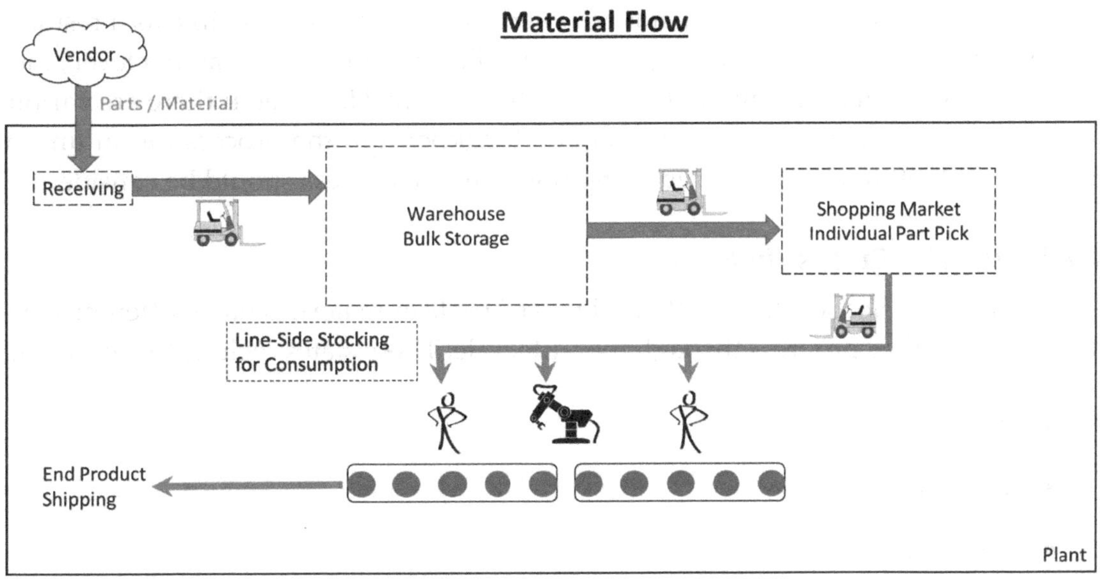

Figure 4-2. Part and material flow through the discrete plant.

3. The plant's Receiving department receives the part. It is now in-plant inventory.
4. The part is transferred from the Receiving department to the plant warehouse, which is considered the *bulk storage area.*
5. When the part inventory in the part pick area (also known as the *shopping market*) drops below a threshold value, parts are transferred from the bulk storage to the shopping market.
6. As the orders are launched, parts are picked in the order in which they are needed by the production line.
7. Parts are then delivered to the production line (in some form of LSS or part feed) and are no longer considered in storage but as *work in process* (WIP).
8. After the final product is manufactured and passes quality checks, the parts are considered consumed and removed from the plant inventory. Typically, this is when the part is backflushed in ERP, and if not already done, the vendor is paid for the parts. This means that all inventory is removed from the plant inventory, and if the parts are on consignment, billing is triggered for payment.

In some manufacturing environments, an order may have multiple reporting points where parts are backflushed or consumed. The number of consumption points is determined by a production management policy and should consider the likelihood of WIP sitting at a given stage of production for a period of time. If the process has no exit points mid-process or if parts are not removed from the production line as a result of nonconformance (thereby spending an uncontrolled amount of time at that level), then there is justification for having only one backflush point. However, if there are various ways that WIP can exit the line or wait at a breakpoint in the process for an uncontrolled length of time, multiple reporting points in the process should be considered.

4.1.5 Material Status (In Plant)

As mentioned in Section 4.1.3, "ERP and SCM," parts or materials have different states in the plant as they progress through their lifecycle. These states include the following:

- Ordered
- Received[1]
- In warehouse
- In shopping market
- WIP
- Consumed

The parts count in each individual state can be tracked to show how many parts or materials are in the plant to support the production rate, as well as to keep an accurate inventory count for financial and tax purposes. This inventory function in MES is also important to ERP in the management of parts and materials throughout the manufacturing process. When the materials or parts are consumed, the backflush function in ERP will be triggered to maintain the proper inventory for the plant and also to determine when the next order for parts should be sent to the vendor. Ordering parts can be automated when the plant inventory drops below a certain threshold.

Parts can be ordered in many ways depending on the usage of the parts in question. For example, if the parts are linked to a specific order and option content, then an advance build notification can be used to order parts. If the parts are not tied to a specific build or option content but are shipped in bulk (like screws), then the parts can be ordered either by tracking the part consumption rate or by whether or not the parts drop below a certain inventory threshold—for example, if a certain screw must be ordered when there are only two boxes of 100 left or if the consumption reaches 200 individual parts used.

---

1. Received, in warehouse, in shopping market, and WIP are all part of in-plant inventory at this stage.

4.1.6 Different Material Management Methods

Now that we have covered the overall lifecycle and business process of the parts and materials in the plant, we must cover the specific processes or methodologies that enable the transfer of parts to the line. This is necessary because these processes and methodologies execute in different ways depending on how parts are replenished on the production line. MES solutions provide a better way to manage these processes. We will cover each one of these in depth:

- E-kanban
- Kitting
- Subassemblies
- Bulk parts
- Inventory control and reordering

Kanban refers to a process where parts inventory is monitored at the line. When the inventory drops below a threshold amount within a kanban container, a manual process is triggered to replenish the line with a predefined quantity of parts. *E-kanban* refers to the use of information technology (IT) to automatically trigger the replenishment process and provide *bin tracking* that is managed by an IT method. This process is useful for parts that have defined storage locations unique to each part or are not limited by end-product option content. Automated ordering and in-plant transfers eliminate the need for a manual process, which may not be triggered at the proper time. An MES benefits kanban by not only ensuring that the reorder process is automated but also by being able to annunciate faults when parts are not delivered or transferred in a timely manner. These faults can trigger the proper response team in the plant to respond immediately.

Kitting refers to putting together a set of parts that has a specific content definition.

There are two types of kitting:

1. Product-specific kitting
2. Production-sequence kitting

Product-specific kitting is the set of parts placed in the kit for a specific product or build job (sometimes referred to as *pegging*). Parts that do not belong on the product will not be placed in the kit. This process helps ensure that the right set of parts for a specific product is delivered to the line, minimizing operator error. For example, the parts for an automotive instrument panel, which can include gauges, radios, environment control modules, and more, can be placed in one kit, and each end part may have different options and, therefore, different parts within the kit.

Another form of kitting, used in the automotive industry, is called *in-sequence production* and is covered in Chapter 3, "Order Execution Management: Details and Benefits." *Production-sequence kitting* (which can be thought of as pegging on steroids) is used in manufacturing environments with mass customization. A kit would contain the parts (usually subassemblies) for a series of end products, with each subassembly slightly customized to a specific configuration. The build sequence for any set of *n* number of products is required to pull a specific part type; the set of parts is then placed into the kit in the sequence specified within the order. For example, consider the seat covers for automotive seats where the kit of *n* seats to be built (say, 5) has the following requirements:

1. Base cloth black
2. Leather brown
3. Upgrade cloth gray
4. Leather gray
5. Base cloth brown

An MES benefits kitting by automating the pick list generation and printing process so that errors in product-specific kitting or production-sequence kitting do not occur due to manual generation issues. An MES also creates a fault annunciation if the kits are not delivered to the line side in the proper time allotted, and the fault can trigger the proper response team in the plant to respond immediately.

A *subassembly* is a set of parts that is assembled before being sent to the main line for assembly on the final product. With material management, the proper parts for a subassembly can be more closely controlled with an initial part pick to make the subassembly and the actual part pick at the subassembly workstations. This is sometimes done to improve build efficiency or for quality assurance so that the subassembly is defect-free before moving on to the final assembly workstation.

A benefit of using an MES is that it facilitates coordinating parts and the end products that require subassemblies. Another benefit is that the MES creates fault annunciations based on the material management process to immediately trigger the proper plant response team.

Bulk parts include all parts that are universally used over the entire range of end products and are not tied to option content. These parts may include screws and other metal and plastic fastening objects or other parts that do not require special handling containers. Because the parts are not linked to option content and, therefore, may not be on every end product, the consumption is not tracked by part pick lists; it must only be tracked by the number of parts used per end product. Once the consumption rate and the amount of inventory required in the plant are determined, part replenishment from the vendor can be automated. This business function is like kanban and has similar MES benefits.

Throughout the whole process, tight inventory management, which includes controls and reordering, must be maintained. With an automated MES material management function, inventory for each part or material becomes more accurate, and automated reordering becomes possible. Similar to stockouts, improper inventory control can starve the production station if a part or material is not available to be processed onto the end product. However, too much inventory is also an issue because the plant has a finite amount of space to store the parts and materials, and, in some places, inventory is a taxable commodity, which means the manufacturer has to pay more taxes because it has more inventory. This taxable expense has no bearing on the efficiency of the manufacturing process and must be minimized.

4.2 Material Management Benefits

To summarize, the benefits provided by using MES in a material management processes include the following:

- Supporting just-in-time production scheduling.
- Minimizing in-plant inventory, which can be a taxable commodity.
- Providing more accurate inventory.
- Leveling inventory to maximize production while minimizing inventory.
- Directly supporting the kitting process by automating it. This also helps minimize parts being improperly selected and therefore avoids end-product quality issues.
- Streamlining the subassembly process to minimize defects before the final assembly process.
- Automating ordering functionality and in-plant transfers. Ordering and transfers no longer require a manual process that may not be triggered at the proper time.

- Giving the ability to track the parts and products through their manufacturing lifecycles.
- Supporting real-time fault annunciation of the different material management faults that negatively affect production, which speeds up resolution time.
- Automatically backflushing parts, which in turn triggers payment. Vendor payments no longer require a manual process, which may not be triggered at the proper time.
- Ensuring that parts, when picked for delivery to the line, are sequenced properly, minimizing the chance that the wrong parts are picked for a specific product build.

5 Quality and Traceability: Details and Benefits

5.1 Introduction to Quality and Traceability

Imagine that you are a customer of a manufactured part, such as a car, a hot tub, or some other product that has a significant price tag. You have had this product for a few months and things are going well. Then, you see in the news that the company that made your product is announcing a recall for a quality problem. A couple of days later, you receive a letter stating the product you purchased is related to this problem. The letter asks you to call your local service depot to arrange for the product to be reworked to fix the issue. If you closely read the letter, it generally says one of two things:

1. Your product is *suspected* to have the problem, and the service center must verify and fix it.
2. Your product *does* have the problem, and the service center knows what to replace or fix. They just need you to schedule the repair/rework.

The difference between these two scenarios is a function of the level of quality assurance that the manufacturing company has engaged in and, more particularly, the level of traceability that is used during manufacturing. Believe it or not, most companies put a lot of effort into ensuring that the material used during production is of good quality and that the product has been manufactured the right way. How *effective* the company is greatly depends on the use of technology and, frequently, on the use of a manufacturing execution system (MES) as a primary component of that technology. In this chapter, we will look at the broad range of quality activities within manufacturing and detail how MES supports those activities.

5.2 Quality Assurance or Quality Control

There was a time when quality was only concerned with answering the question, "Was it built correctly?" Manufacturers engaged in activities during or after the manufacturing process to verify the product *was* built correctly, or they helped establish processes during manufacturing to ensure it *is going to be* built correctly. The scope of quality has expanded considerably since then, but as far as the manufacturing floor is concerned, these questions are still the primary concern. The field of quality includes functions such as various types of inspection, functional or performance testing, statistical process control (SPC), and continuous improvement activities. It also includes quality functions on the production floor such as equipment calibration, user certification, and documentation control for work instructions. All these activities can still be brought back to the original two questions: (1) Was the product built correctly (frequently referred to as *quality control*)? (2) Is the process understood enough to ensure that the product will be built correctly (also referred to as *quality assurance*)?

If the Quality Control team determines that the product has not been built correctly (which is referred to as a *nonconformance*), then steps are required to repair the product or scrap the nonconforming product and replace it. In some industries, the nonconforming product may also be downgraded to a lower level of acceptance (a product of lower quality), but for all intents and purposes, it is still being removed from the order as part of the intended product quantity. Downgrades are simply a means of recovering the costs of scrapped products.

Part of the quality assurance (QA) team's role is to analyze the failures and nonconformances that occur to determine how to prevent them from happening again, all as part of a program of continuous improvement. There has been a considerable amount of growth in this area with the development of methodologies like Lean Manufacturing (Lean) and Six Sigma. The quality function of continuous improvement and MES is an expansive subject that will be covered in greater detail in Chapter 7, "Continuous Improvement: Details and Benefits."

Throughout this chapter, we will discuss the issues that are related to the activities just described and how MES has helped improve the field of quality on the factory floor. Finally, we will discuss traceability and the extended benefits of using an MES for traceability.

5.3 Quality Assurance

The driving force behind QA is that it is far more reliable to prevent problems from happening than it is to detect and correct them. As a result, most of the quality activities have moved toward preventing defects instead of trying to efficiently detect and fix them. In fact, many companies have realized that trying to improve the repair process to put products back into mainstream production quickly can be detrimental to process improvement because it removes the opportunity to collect data to determine the root cause of the defect. The only time that defects should be repaired and moved back into production quickly is when the root cause of the issue is already known, and the extra data collection can provide no additional insight.

In the next few sections, we will cover many aspects of QA, the issues that Operations faces during the production process, and the way MES has helped resolve those issues.

5.3.1 Calibration of Tools and Equipment

One key facet of QA is to ensure that equipment and tools are in a known good state of operability. In addition, Operations must be aware of the most up-to-date performance characteristics of the equipment and tools. In most cases, the quality team will highlight and audit the need for implementing a management program for calibration or preventive maintenance, and the maintenance activities will be carried out by Plant Maintenance or Manufacturing Engineering staff.

As part of a typical quality management program, the Quality team is frequently the owner of the calibration lab with support from Manufacturing Engineering. The program keeps test and manufacturing equipment in a state of known calibration compared to national standards and establishes calibration policies and procedures. To do so, it is necessary to maintain a record of when tools and equipment require calibration (or maintenance) and where the equipment is located on the production floor at that time. Large pieces of equipment are relatively easy to track, but small tools and pieces of equipment are often moved by Operations personnel to where they are needed, making them more difficult to track. In some cases, the whereabouts of the tools or pieces of equipment are not known until someone uses them in production and discovers that the calibration was due earlier in the week.

Many production areas will establish prestart checklists to ensure that issues, such as missing equipment, are detected. However, as these are manual checks, sometimes checklist items are missed. Having an MES can help track the equipment and tools used during production and compare the calibration records with the current date (or utilization quantity or another record of activity) to verify that a specific piece of equipment is within its accepted calibration period. If the equipment is out of the calibration period, the system can prevent the operation from moving forward until something is changed. As part of the production operator's initial start-up, the checklist can include scanning the tool's tracking number into the MES to verify it. Additionally, because the system is aware of the calibration due date and the location of the equipment, it can send an alert to Manufacturing Engineering that the equipment will soon be due and must be replaced.

Most MES can interface with a company's calibration/preventive maintenance system to keep an up-to-date record of when calibrations have been performed and, more importantly, when they are due. Because a considerable amount of functionality is needed for a calibration management system, the management of the actual work being performed during calibration typically is not within the MES scope of functionality.

Another part of the calibration program is to determine how often calibration is required. In most situations, the company will start with a relatively conservative time of three months (or something of this nature). However, as the history of the equipment develops and adjustments are not needed during calibration reviews, the calibration period can be extended. The problem with this scenario is that there is no specific reason for choosing these time periods, and the times are extended until a determination is made that the equipment must be adjusted. By that time, the equipment could have been in use while it was in an *out-of-spec* condition and could be a root cause of a quality issue. If this is correct, then what is the real answer? How often is calibration required?

With most pieces of equipment or tools, usage causes degradation in performance. The issue is that without automated tracking, it is difficult to track equipment or tool usage in most production environments. Using an MES, all production usage can be recorded, and the data can be used to determine when maintenance or calibration is required. In addition, because the equipment or tool usage is directly linked to the products it was used with and possibly the resulting nonconformance, Quality can justify adjusting the calibration/maintenance period as required. In a more advanced application of MES capabilities, after a performance characterization can be determined, the calibration request can be determined by sensors within the equipment or tools that indicate a performance degradation.

Calibration and preventive maintenance management share all the aforementioned issues and possible solution support. Advanced analytics for monitoring tools and equipment, as well as other aspects of maintenance management, will be covered in more detail in Chapter 6, "Production Monitoring: Details and Benefits."

5.3.2 User Certification Management

User certification and calibration have a similar concern: when (and on what piece of equipment or tool) was the user certified? Most user certifications are set up to ensure that an operator has received training to perform specific production operations. A general certification may be established to operate specific equipment, including the setup and safe operating procedures. If a specific procedure is unique enough for a product or if the consequences are critical enough, a specific certification may be required for a particular piece of equipment/product combination. A quality management decision to create a requirement for certification is based on the level of uniqueness of the work and the criticality of failure of adequate performance. However, in some cases, certification can also be linked directly to regulatory requirements, as in pharmaceuticals. In this case, if the certification has expired or was not managed, the product would not be available for sale.

The complexity of certification management also must be recognized. Having every task on the production floor be a *certified task* is of no value if the labor needed to maintain the certification records will cost more than the quality issues being addressed. As in all things, the cost of action cannot exceed the benefits derived from that action; this is a fundamental rule of good operations management. So how can user certification be managed cost-effectively? The answer lies in automation and standardization.

By automating the validation of user certification, the validation is configured into the day-to-day procedures of any operation. By interfacing the MES with the user certification management system, the MES will have a record of who has been certified and for what aspect of production the user has been certified. Most MES can be configured with unique identification for any product, equipment (resource), operation (if different than the resource), and tool. If the MES has been configured correctly, it can validate a certification for any combination of these data elements (sometimes called *data objects*) and distinguish changes in real time, making it easy for the system to keep up with whatever variation of production throughput is required.

The other aspect of managing certification is to standardize as much of the manufacturing process and the way information is presented during production as possible. This enables a more generalized certification scope. The more standardized the procedures of operations and user interfaces, the less unique any operator task will be, resulting in less need for specific user certifications. Additionally, if solutions are created properly, standardized interfacing and user interface functionality will significantly benefit the setup and support of user configuration requirements.

Earlier we mentioned the cost of protecting against an issue. Estimating the cost of an issue is not easy. It requires the following:

- A detailed understanding of the processes involved in the issue
- The labor that is lost as a result of creating a product with a nonconformance (i.e., the direct production labor time that is lost)
- The material that will be scrapped as a result of the nonconformance, as well as the extra time and material needed to analyze the nonconformance
- A decision on the rework/repair option and then performing the rework or repair (or scrapping the material)

In addition to these items, the frequency of the issue must be estimated to understand the full cost of a quality issue. In a non-MES environment, this estimation will be at a rough order of magnitude and, therefore, will have a considerable margin of error. As part of the cost analysis, the manufacturing or quality engineer must request manual data collection from the production staff, which will inevitably cause slowdowns in production and possible errors as the staff members are distracted from performing the work they normally do.

Using a well-configured MES, combined with properly defined processes, the accuracy of the estimate can be greatly improved. This is not to say that the costs will be exact (after all, it is an estimate), but the accuracy of the estimate will be orders of magnitude better than in a non-MES environment and can be fully justified in the data used within the calculations.

5.3.3 Poka-Yoke Functions and the User Experience

After a process has been running for a short time, the Manufacturing Engineering staff will start to understand its characteristics. As part of that understanding, they will gain insight as to where in the process production operators are having issues with the repeatability of that operation. Issues with repeatability may include missing steps within an operation and/or an inability to maintain repeatable cycle times. This is where the use of poka-yoke methods comes into play.

Poka-yoke (or *error proofing*) is the Japanese term for adding a step or device to prevent a specific problem within an operation. Poka-yoke methods have existed in Operations for some time, and anyone familiar with the current quality management methodology will understand them. This section focuses on applying poka-yoke methods to MES.

Poka-yoke methods in MES are incorporated by presenting information in the way of reports and work instructions (WIs) and in designing the user interface (UI) to properly sequence the data entry and procedure for the production user. One example of the poka-yoke method includes sequentially stepping through which data entry is needed and then validating the format as data is being entered. If there is an error in the format or the range of the value, the operator is alerted. Another example is using *pick-to-light* interfaces (shown in Figure 5-1) that guide the operator to select the material from a line-side stocking (LSS) bin and then detect that the operator is actually picking that part, again validating for errors in real time.

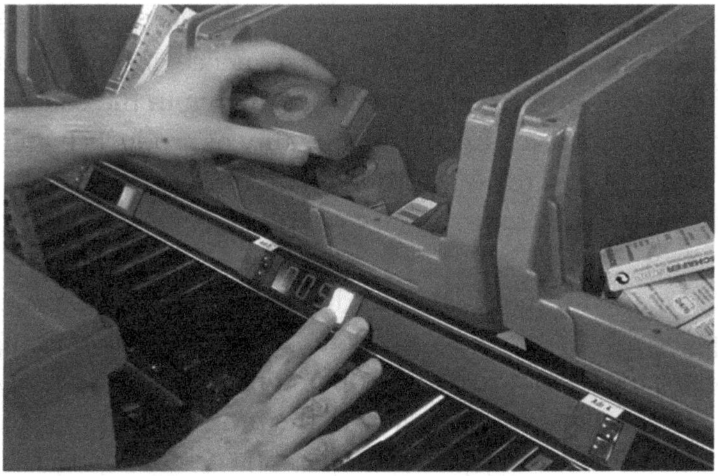

Figure 5-1. Pick-to-light interfacing.

Pick-to-light interfaces can also be beneficial to material management, which was covered in Chapter 4, "Material Management: Details and Benefits." If the pick-to-light interface is not possible, using barcodes to scan the picked material is an alternative. In this case, at least the correct genealogy is recorded. However, this method requires manual processes to replace the source barcode (scan card or container label) when the material is replenished. Barcode scanner technology has increased considerably in recent years to provide a solution that can be used for just about any application.

Implementing these methods into an MES requires understanding the steps to complete an operation (from the operator's perspective) and understanding how to set up the data and user experience according to those steps. Within the flow of the user experience at an operation, the UI can lead the production operator through the steps (preventing the next step if the current step is not completed), or data collection (genealogy and/or parametric) can be made unique enough for the system to determine which data entry requirement is being met by the item scanned or by the format of the data value entered.

Having the system use data to determine what steps are completed—versus providing the steps for the operator to follow—can simplify data collection, but it enables the operator to determine the step sequence, which in many cases may be detrimental to the process and in most production scenarios is very difficult, if not impossible, to correctly determine all the time.

Leading the production operator through the correct sequence of steps by controlling when data can be entered and automating that data entry can be simpler to implement. Getting the screen flow to correctly support the production operator, however, requires a good understanding of not only the sequence of steps for the operator but also the flow of data within the system that is occurring in the background. The benefits of repeatable operation steps, consistent quality of products being manufactured, and clean data entry in regard to that operation, however, are well worth the time spent to properly design the production user experience.

In some production environments, the production operator is expected to be able to perform the sequence of steps for a given operation with as little training as possible. In these environments, the UI for production operators must be simple and provide clear, human-readable instructions for each step in the procedure. As long as the sequence for that operation (and even every product/operation combination) remains consistent for a long period, it is well worth creating a detailed set of WIs for that procedure and tying them into the MES screen interface, and then ensuring that the UI is updated with each step in the procedure to match the WIs. Most MES can present a sequence of specific WIs in precisely the correct order to match the procedural steps by product, operation, and resource combinations.

In these situations, it is also very important to understand the nature of maintaining the WIs and UIs during the MES implementation phase. Many MES implementations have gone off track because the maintenance of the supporting systems (including the MES itself) was not properly understood at the time of implementation.

In other production environments, production operators are highly trained and have a broad knowledge of the production process and the equipment they must use. With skilled production operators, the UI can become quite complex and can be used to provide a lot of information on one screen. For these kinds of users, it is important to balance the complexity of the UI with the practical aspect of interpreting the content on the screen. In addition, it is important to provide an intuitive exchange of information and data input to give the operator access to the most information with the least number of keystrokes or button clicks. When considering the design of solutions for users at this level, it is also important to standardize the way data is presented on screen. This requires a well-defined *look/feel* policy and normalized terminology as part of the implementation process. Taking the time to design the user experience will help in the long run.

In general, solution design for the MES environment is different from any other enterprise system because of the detail in which the system must mimic the production floor. Care must be taken in solution design to recognize not only that level of detail but also a general operations mentality that is prominent within the production manufacturing environment. In manufacturing, there is a strong expectation for the efficient flow of product and material—an expectation that production operators learn from their first day on the job. This expectation must be reflected in the user experience as operators interact with the MES. The system (1) cannot slow the operator down by requiring too many actions to input data or too many screens to obtain information, and (2) the solution architect must recognize that the solution design must be intuitive for the operator's next step according to the real-time status of the UI sequence. Systems (and designed solutions) that do not deliver in these areas will not be accepted in production, and production operators will search for (and likely identify) ways to get around the system. The importance of an effective design of the user experience in MES should never be underestimated or under delivered.

Systems that provide the capability to reconfigure a UI without having to recode the development of that UI will be valuable in the maintenance and support of an MES environment as the use of the system matures. MES applications that provide a high amount of configuration for the UI are very valuable for the flexible support of changes to the user experience, and they enable the Manufacturing Engineering staff to configure user screens themselves. However, that highly configurable capability requires a strong understanding of the data manipulation by the system in the background.

As the technology used in manufacturing grows (as a result of concepts for the Industrial Internet of Things—IIoT), there will also be a growing need to develop specific interfaces to tools and other systems that may require small blocks of specific coding to pick up or deliver data automatically and incorporate that data into the UI.

As part of the requirement for MES, companies should also look for modular interface capabilities that can be included in the UI configuration capability.

5.3.4 Process Monitoring and SPC

The aspects of an MES that make it such a value to manufacturing are (1) the real-time processing of data and information and (2) the correlation of all data to particular actions taken within production (e.g., the assembly of a component to the parent assembly or the inspection of a particular product). Using the correlations to actions within production makes data mining to determine a root cause much easier. After data has been collected, the system can immediately make that data available to any other operation or procedure step on the production floor. As a result, for each serial number of a parent assembly, an MES can record the operator who performed the assembly and the workstation that was used. If the configuration of the production environment is detailed enough, the MES will also record the serial number (or batch ID) of each component that was assembled, the tools (with tool IDs) that were used during the assembly of each component or material (e.g., if a powered screwdriver was used), the number of turns used to torque the screws down for each screw that was assembled, and the end torque that was applied to each screw. This data is recorded in real time for the assembly of each screw for each product. Other tools may have different performance characteristics that can be recorded at the time an action is performed and reflect on the real-time aspect. Each action at an operation can be validated and controlled and can interact with other equipment or interfaces (e.g., possible vision systems) as work is performed.

Continuing with the previous example, this data then becomes available to report or act on immediately after each assembly of screws. Therefore, if the torque of one of the screws is out of specification, the MES can immediately (before the next screw is assembled) alert the operator about the problem, stop the assembly of other screws, and record a nonconformance against the parent assembly. Depending on its configuration, the MES can also stop other operators from performing the next screw assembly if it detects a sequence of failed attempts (e.g., a bad batch of screws) by initiating a quality event to prevent the assembly of more products with that batch of screws. Using the assembly records of previous products made with screws that have the same batch ID, the system can also place a hold on all products anywhere within the process to stop the completion of additional assemblies that might make repairing the problem more difficult. All this activity can be initiated within seconds of detecting the problem.

Whether a production manager would want production to be stopped like that depends on the management policy on quality issues. The main point is that MES can provide the data to make decisions like that and can react quickly and broadly to production issues. These types of scenarios must be reviewed and become part of the configuration for the MES implementation. If configured correctly, these scenarios can be enabled or disabled as required, but they should be discussed and designed properly as part of the initial implementation.

Whether you are tracking powered screwdrivers or other powered/automated tools used in production, the same aspect of data recording, validation, and subsequent control is an inherent part of MES operation.

After the capability to monitor a process has been established, the next step of the monitoring activity is implementing an SPC program to track the conditions of a process and initiate alerts and/or controls when the process shows signs of going *out of control*. With all processes, a certain amount of variation is expected. An SPC program is implemented to monitor a process when the key parameters are known and can be measured, and to alert the production operator when it recognizes variation that is not considered normal. The issue with implementing SPC is the need for the operator to both perform the work required to manufacture the product and record the details of the process key parameters. While the key parameters are recorded (and plotted on a graph for easy interpretation), the production operator is also required to look for patterns in the process variation. This is a lot of work for an operator who is also trying to keep up the product flow through production. As a result, without automation like MES, SPC is frequently implemented half-heartedly and patterns in process variation are often missed. Most MES are capable of receiving key process-characteristic data as part of the normal data collection steps to support SPC.

As shown in Figure 5-2, the MES collects and monitors key parameter data from the manufacturing line. This data is then automatically plotted onto a graph that can be constantly displayed as part of a *billboard* display system. Alternatively, the data is recorded and monitored in the background, and a graph is displayed for the operator only when specific patterns in the process characteristics are identified or on request. Either way, the MES monitors the key parameters and identifies patterns, enabling the operator to focus on the production work.

After an *out-of-control* event has happened, the process is stopped, and the cause is determined. Manufacturing Engineering then makes any required adjustments and enables production to continue. The SPC module in an MES can be configured to intervene in the operator's work and stop production if an issue is determined to be critical. This aspect of SPC is frequently overlooked or ignored by production managers, but it should be an important part of the MES implementation.

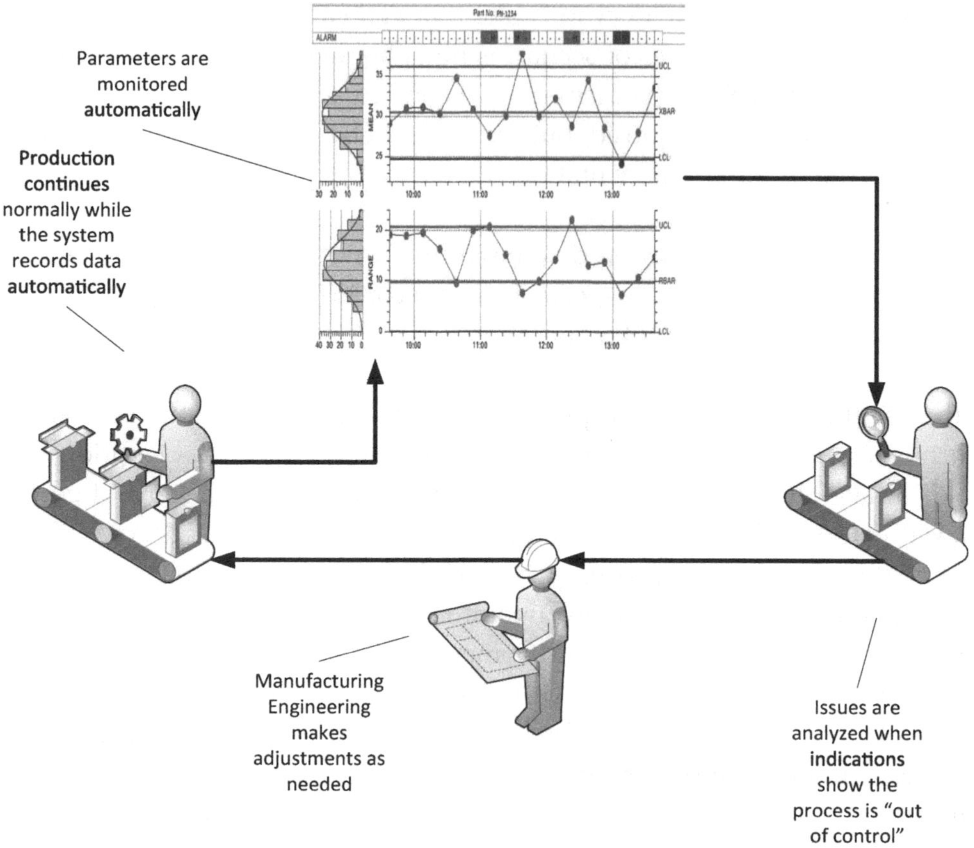

Figure 5-2. Using SPC charts with MES.

The nature of SPC implementations is to provide immediate notification and intervention at the time the problem is detected. Therefore, it is a short-term *monitor-and-correct* solution when the characteristics of a process are known.

5.4 Quality Control

As stated earlier in this chapter, the function of quality control is to ensure that a product that has been manufactured (or is being manufactured) meets performance or functional expectations. As a part of quality control, processes should contain control points (points in the process where "go/nogo" decisions can be made) that are used to verify product characteristics before moving on to the next stage of production. Control points in a process are usually selected when a work in process (WIP) has reached a *point of no return*; in other words, if there is a problem with the product characteristics, it must be fixed at that time or it will be difficult (or impossible) to fix later in the process. This would be the result of critical materials or components no longer being available or additional damage or cost that could be incurred if the problem were detected beyond that point in the process.

At these control points, quality and manufacturing engineers will determine the type of control intervention (inspection or testing) that best suits the scenario. This is determined by the aspect of product characterization that must be detected at that point in the process. Product characterizations of an aesthetic nature (the validation of components being present or the presence of aesthetic damage or some other visually detectable characterization) would be part of an inspection step, whereas validation of performance or functional capability would be an aspect of testing.

After the quality control point and the method of quality control are defined, the manufacturing engineer must determine how to identify these failures at that control point and how to handle them, all as part of a nonconformance management process. As part of the control definition, the manufacturing engineer should keep the focus of control on detecting particular issues. Having the focus of control defined too broadly will make the detection of a particular issue more difficult and much more time-consuming. The issues that are defined should be specified to detect an issue only until it has been eliminated by a process improvement change. After the issue is proven to be corrected by a process change, the focus of that control point should then be changed to the next greatest issue (as is reflected in the tracking of issues from Pareto charts).

As shown in Figure 5-3, the data collected via inspections and/or testing should be reviewed at a minimum from a Pareto context. The details of what is reported depend highly on product/process combinations.

Using the Pareto chart as a reference, identify the most significant problem and focus on the control points required to determine the cause of the problem. (This may include temporarily changing the data collection parameters of the process to collect specific parametric and nonconformance data.)

Once the cause is determined, make changes to the process or product to eliminate the problem. Reinspect or test the process to ensure the problem has been corrected and compare the results with data from previous production runs. Then, evaluate the next most significant issue. Remembering the Pareto rule (20% of errors are causing 80% of fallout) will keep the continuous improvement on issues that have the greatest impact. In many cases, this rule should also be reconciled with guidance from the Theory of Constraints (i.e., an improvement on a nonconstraint does not improve throughput). The data provided by the MES and changes in the data collection procedure to support these methodologies will improve a manufacturing environment quickly and sustainably.

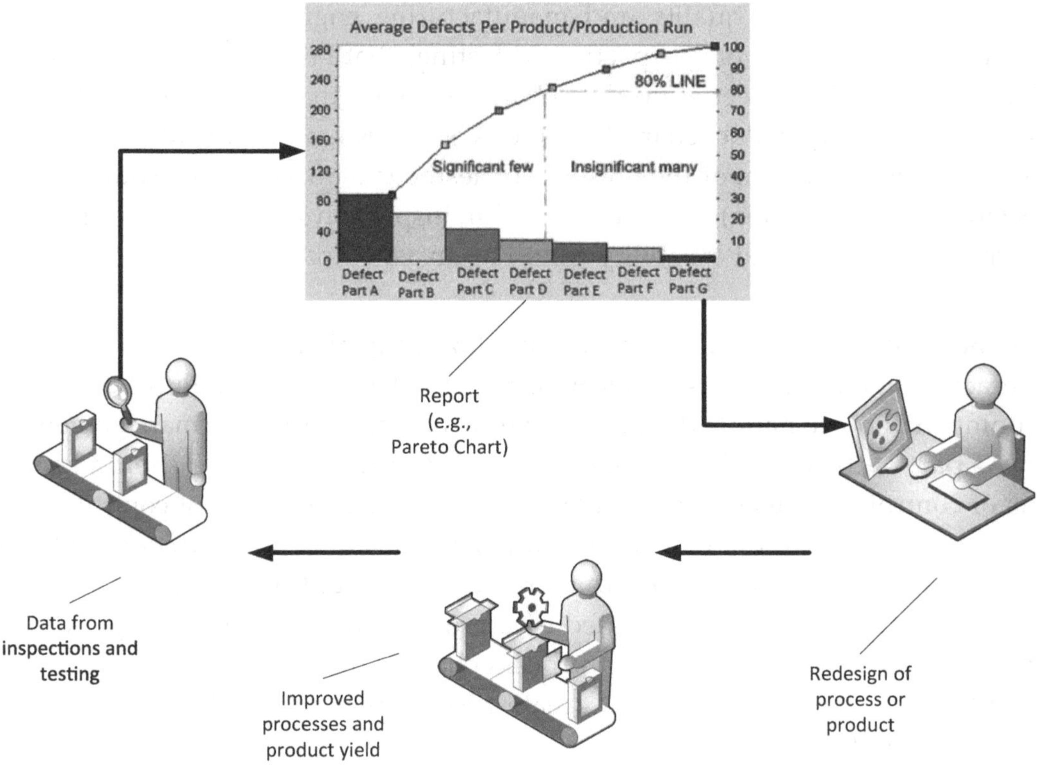

Figure 5-3. Pareto-driven continuous improvement.

The nature of Pareto charts is to monitor trending characteristics over an extended period. As such, they make continuous improvement possible over the long term and are used to determine the key problems in a process.

Planning a quality control program can seem conceptually easy. However, failure to set up and implement a detailed control program can cripple a company's continuous improvement process, not to mention make warranty management extremely difficult.

A full discussion of the aspects of a high-quality control program is beyond the scope of this book; however, criteria that should be considered within the program include the following:

- **Nonconformance identification** – This includes not just how a nonconformance is marked but also how it is described. Care should be taken to differentiate between identifying a nonconformance defect and a symptom of the nonconformance. Symptoms describe how a nonconformance is recognized, whereas defects describe the actual problem causing the symptoms.
  - For example, a symptom is a noisy electric motor, whereas the defect is a cracked bearing in the motor.

    - Identifying symptoms as defects can cause confusion in the analysis process, make defect Pareto reporting complex (which can significantly hamper continuous improvements), and hide critical data as a result of complicating the correlation of nonconformance data.
    - Providing a means of marking a unit of WIP as a nonconforming product will depend highly on the product itself. Tags or folders that move with the product are used frequently. These are called *travelers*.
- **Nonconformance dispositioning** – This defines what to do with the product after a nonconformance has been identified. It includes changing the status of the units that are listed as nonconforming, defining the subprocess that will be used to analyze the problem, repairing/reworking the units, final dispositioning to return them to production in one manner or another, and any changes in status (i.e., returning the units to an active WIP or scrapping them).
    - Contributing to the dispositioning process are the data collection requirements that are needed to feed continuous improvement.
    - A side issue to nonconformance dispositioning is ensuring the processing (analysis, repair, and return/scrap) takes place in a timely manner. A frequent issue with disposition management is ensuring older products are repaired and returned before they become obsolete and must be scrapped anyway.
    - Nonconformances that already have considerable characterization data can be returned to production quickly. Nonconformances that are being investigated may require additional data collection elements to properly characterize the nonconformance itself or to determine its cause.
- **Reporting and alerts** – Part of any quality control program is to report on the issues that are found. In many quality control programs, the problem with reporting is getting information from the reports in a time frame that is of value. Whether it is a report on the trends that production is facing or a particular problem occurring at any particular time, getting the data needed for the reports or getting the information from the reports in a timely manner is always a problem.
    - Reports that present trend information will usually have visibility of timeframes down to the production shift level and should be reported from a database that has *flattened* the data from an operational database (additional details can be found in Chapter 9, "MES Architectural Concepts").
    - Alerts should be characterized from the operational database to ensure the detection of alert criteria in real time.

The next section reviews the way an MES program can help resolve these issues and provide support for any quality program.

5.4.1 Nonconformance Identification

Identifying the problem (the nature of the nonconformance) and the material or component that contains the problem seems like an easy issue to solve. But how do you mark the WIP as having a nonconformance so that it does not get mixed up with good products, and how do you describe what is wrong in a manner that is clear and easy to understand without a long explanation? Many companies use a traveler or a tag that is attached to the WIP unit as an identifier. But what happens if the traveler or tag is separated from the unit or destroyed? Now you have a situation where nobody knows what is wrong with a nonconforming product unless it is reinspected or retested. This is an all-too-common scenario.

The key reason for using MES functionality in the form of eTags or eTravelers to record and maintain the data for nonconformances is that unless the data is explicitly deleted (which best-of-breed MES will not allow), nonconformance data that has been recorded and correlated to the serial number (or batch ID) will remain with that serial number for the rest of its life. In other words, as long as the WIP identity remains within the database, the record of the nonconformance (and the action taken to resolve it) will as well. The point is that even if there are multiple nonconformances (or a resurfacing of an old nonconformance), the complete history of that WIP unit is always available. If actions were taken to resolve a nonconformance, but the nonconformance was not actually resolved (which is sometimes called a *false fix*), the MES will (1) maintain that record to be used for various reporting requirements and (2) provide details enabling that nonconformance issue to be characterized and to create an alert in support of continuous improvement management. In some industries where the nonconformance is a contamination of some sort (which occurs in chemical and food production), the MES will also contain data about the preceding batch of material and cleaning procedures to determine the source of contamination.

One of the main rebuttals to using an MES for nonconformance tracking is a request from production management for people to be able to read everything. This becomes a primary driver of the tags or travelers in the first place, enabling a person to see the status of a unit just by looking at it. This, unfortunately, takes us back to the original problem of *human-readable* tags being prone to error, such as lost tags and incomplete updates, because the work was completed, and the tag was not updated. Also, it is difficult to conduct an analysis of multiple WIP units from the data on a tag. As long as a person can quickly determine the status of a WIP unit that may be *out of place*, there is little desire to follow up on why the unit is out of place.

Not using an MES to identify a nonconforming unit also propagates the illusion that the automated system is not reliable because staff will continue to rely on the visual system and ignore the data discrepancy (between the system and the tag) as long as there is something that gives direction, even if the direction from the visual system is wrong. Companies that maintain a dependency on visual systems are, in essence, *shooting themselves in the proverbial foot* and ignoring information that can support real growth in production and quality capability.

The next part of nonconformance identification is entering the data clearly and avoiding creating records that contain redundant data elements. (Now you can say, "Can you explain that please?")

The following questions should be answered in a nonconformance record:

- How was the defect detected? (This helps validate the nonconformance or arrange the presentation of the nonconformance (NC) codes within a user display.)
- What is the actual defect? (This is usually an NC code.)
- Where is the defect located (i.e., where on the product or in what batch of product), including the component or subassembly (in discrete manufacturing)?
- At what point in the process was the defect detected? (This helps identify process failures that created the defect and other operations that should have detected the nonconformance.)

However, all this data does not have to be within the NC code itself. The code should only be used to describe the failure and/or defect. The serial number or batch ID found to be defective, the mode of detection, and where in the process the nonconformance was detected should all be included as part of the nonconformance record but not part of the nonconformance coding.

Some companies also want to include "the root cause operation," but experience indicates that until Manufacturing or Quality engineering has completed an analysis, identifying the root cause operation during the production cycle is most likely going to be misleading. For example, a scratch or other damage to the product may not have occurred at the operation where material was added or processed; it may have occurred later in the production process as a result of some other method of product handling. Until Manufacturing Engineering analyzes the occurrence, the part of the process where the defect actually happened is not known for sure. The data correlation for nonconformance issues offered by an MES will provide significant guidance on the root cause and should be the first point of reference for manufacturing and quality engineers.

5.4.2 Nonconformance Dispositioning

In the previous section, we reviewed issues to keep in mind when examining nonconformance identification and how an MES can help with the identification and management of nonconforming production units. The next step in the nonconformance cycle would be to take action to resolve the nonconformance.

The dispositioning of a production unit must serve two functions: (1) separate the nonconforming unit from the mainstream process to ensure that it is not confused with regular production, and (2) ensure the unit is reviewed and analyzed to validate the nonconformance and determine a course of action to resolve it. A third function that should also be a part of the dispositioning process, and is often overlooked, is to provide data that can be used to determine the root cause of the nonconformance as mentioned in the previous section.

In a non-MES environment, after the inspection or testing has been performed, the nonconforming unit would be tagged (to record what was wrong with it) and placed in a container, which would then be quarantined (in a QA-restricted or repair area). In the manufacturing of complex equipment, the nonconformance would be isolated to the smallest subcomponent, and that component would be replaced and quarantined in a similar method. The quarantined units would then wait for someone to evaluate them. At this point, there is a substantial difference in the way quarantined units are handled depending on the industry and on the repair process key performance indicators (KPIs)—if they are measured. However, the outcome of each process would be very similar.

The primary issue with this scenario is that there is frequently little guidance or control over the process of repairing units and returning them to production. If there is some guidance from performance measurements, the KPI used is typically based on the volume of units repaired and returned to production. This kind of environment will result in little to no evaluation to identify the root cause of the nonconformance. The emphasis is on fixing and returning more units, so no assessment of the process or material that caused the nonconformance is engaged. Further complicating this issue is the fact that repair operators are then encouraged to spend time fixing units when the repair action is already known and no diagnostic effort is required. The units that do not have a known solution will sit on the shelf until someone finally decides to scrap them, and any information that could lead to a process improvement (which could improve throughput and production capacity) is lost with them.

If there is an attempt to identify the specific issue that fixes the nonconformance (not to be confused with identifying the root cause), the repair operator will replace different components in an effort to identify the defective component. In a non-MES environment, this results in no history of any previous repair efforts. Each push to identify the problem must start by performing the same initial analysis, or the repair operator must try to remember any previous efforts that were taken. In some cases, the problem may have been resolved once before by a previous repair operator on a previous unit, but that information was never transferred.

Because there is no process-level data to analyze, which might provide guidance on process improvement, there is little incentive to perform such evaluations. This lack of analysis is a prime cause of most companies having an almost nonexistent continuous improvement (CI) program and leaves Quality staff with little reason to manage any kind of effort for improving processes.

Having an MES can significantly change this scenario both in the processing of the nonconformance back into production and in support of the root cause analysis for the Quality CI program. More information on root cause analysis is provided in Chapter 7.

MES Support of the Repair Operator

If any components have been replaced on a unit in relation to a nonconformance, a well-configured MES will have a record of it. Depending on the MES application and the configuration of that implementation, the MES may also be able to indicate whether the repair/rework was successful. This kind of history can provide guidance on future nonconformances or input on what has been unsuccessfully examined so far for a current nonconformance.

In addition to guiding repair operators on how to fix a problem, the MES can track the analytical skills of a repair operator. A senior repair operator or engineer can review the repair history related to a more junior operator and determine if the efforts may have been misguided or particularly effective. This information can be beneficial in training repair staff. A beneficial KPI that can be used in this case is the number of attempts made to identify the issue that was the problem. To support this KPI, it would be instrumental to have the MES report on the actions (and the number of actions) taken by the repair operator that resulted in a confirmed repair.

MES Support of Functional Testing

Identifying the root cause when a unit fails testing can be a little more difficult. The primary issue with test failures is that the test results only describe a symptom of the problem, not the actual nonconformance. The results describe that a test (or series of tests) failed and provide the test parameters that identified a failure. But the results do not tell you what is wrong or what is defective. To get this information, the WIP unit must go through additional analysis to determine *why* it failed.

Depending on the testing sequence, determining which tests failed can be very important when identifying which component failed. In most cases of functional testing, a specific series of tests fail due to a cause of failure. The series of test failures can then be used as a signature of what has caused the failure. When test results are viewed in this manner, having a record of which tests failed and the actions that may have been taken previously to correct the failure can significantly help the repair person quickly pinpoint the problem. Having the MES record this information provides the correlation needed to identify the signature of a test failure, and that data can be used in various reporting schemes (Pareto charts, repair histories, and UI alerts) to guide the repair operator through the repair process.

When using test failures as part of the nonconformance reporting structure, it is important to recognize that the test(s) that failed (reported by test IDs) simply provide additional data related to the failure. Because multiple test IDs can be related to a specific failure cause, the test IDs should not be used as the primary branch of nonconformance reporting.

Nonconformance Routing

Determining the routing for nonconformance handling is highly dependent on the product, industry, means of nonconformance detection, and skill set of the manufacturing floor. As a result, defining specific scenarios of best practices for an MES to assist in directing products through the nonconformance steps would be impractical. However, this section will cover some key points that should be assessed when looking at an MES and some manufacturing issues that MES solves for processing nonconformance.

One feature of most MES that can be valuable is the system's ability to automatically direct nonconforming units to a series of repair operations. When the system is made aware of the results of tests or inspections, decision points can be added to the production routing that will recognize that a unit has a nonconformance record. Depending on the complexity of the repair requirements, the system can analyze the details of the nonconformance record (e.g., the product, type of nonconformance, and other factors the system is aware of) and determine the difference between multiple repair operations to fix specific problems. If the nature of the nonconformance is already understood, the system can direct the unit to a quick repair operation; if no details are available, the system can direct the unit toward a more significant repair and analysis operation. When enough data is available to characterize a problem, having the system make these kinds of decisions is possible.

Common problems with implementing many quality control methods include consistently identifying nonconformances, tracking the nonconforming products through the analysis and repair process, and collecting meaningful data on the nonconformances to be used for CI programs. Many companies struggle with various aspects of these problems.

Having the detailed data collected by an MES gives the Quality team the capability to manage a quality control program and is a major benefit of an MES implementation. This aspect of quality control should be a significant part of the MES program scope.

5.4.3 Testing and Inspection

The next question that should be asked is, "How much should be inspected or tested?" In other words, what percentage of the production flow should be evaluated? The automatic assumption is 100%. Depending on the scenario, statistical techniques can be used to determine what is needed. In a discrete manufacturing environment, this will be determined by the flow rate, the total volume of products being made during a single production run, and the expected fallout rate of production.

First, note that products are tested and inspected only because production is not trusted to make products correctly in the first place. In other words, if the process of making the product was reliable, there would be no need to test the product. This is what the foundation of manufacturing to Six Sigma levels is based on; if production can *prove* that the processes are reliable (i.e., the product can be manufactured without nonconformances), then the amount of testing (or inspection) can be reduced to audit levels and no longer require 100% testing. To have that proof of capability, data is needed. This is a key benefit of an MES because it collects, reports on, and controls processes according to the specific conditions of data in real time. If there is a record of the processes being in control, an MES can direct product for testing or inspection at audit levels, and the system can make a random selection of products to audit. If there is an event (i.e., a sudden occurrence of significant nonconformances), the system can automatically switch to full 100% testing (or to a higher level of auditing) and maintain that new status until conditions return to acceptable levels.

As part of the audit process, an MES can group products into specific identities (process lots) to be controlled within the audit process and can manage or release the group depending on the group's audit results.

When configured properly, an MES becomes a central source of data required to characterize a process. When it is possible to characterize a process, an MES can use that data and the characterizations to automatically make decisions in routing products and signaling alerts.

5.4.4 Six Sigma and Lean

The Six Sigma and Lean initiatives use scientific methods to determine the cost of *process loss*; evaluate the options for controlling that process loss, based on the cumulative cost of each option; and then verify the results after an option is selected and implemented. The costs calculated by Six Sigma and Lean practitioners are determined by performing studies and sample runs on the processes being evaluated. Most of the data practitioners are trying to obtain through these studies is already available through the MES. In addition, because the MES also contains the historic data, the outcome of these project initiatives can be directly compared to previous actual runs, which fits directly into the Six Sigma methods. Many Lean and Six Sigma practitioners have already recognized the importance and benefits of MES.

5.5 Traceability

A scenario was presented at the beginning of this chapter regarding two examples related to the levels of traceability used in manufacturing. In one example, the company would have a record of exactly when material was used and in which products. If a product has a material issue, the company simply examines the records and determines which serial numbers (or batch IDs) were made with that material. As a result of warranty information submitted by customers (remember those warranty cards you get), the company may even know exactly which customers received those serial numbers. In the other example, the company has an estimate of when a particular batch of material was used during production, and as a result, the company could only estimate what products were made during that time and then which products *may have* been made with a defective part. Because of the supplier contract, the likely outcome is that the manufacturer of the end product would probably have to bear the cost of evaluating each product that *may have* been made with defective material. As a result, the original equipment manufacturer (OEM) must bear the brunt of a loss in customer confidence when issues arise. Rarely is the material supplier named for a product recall until well after a public outcry has made front-page news. The public will remember the name of the end-product manufacturer for many years to come, but few will remember the name of the material supplier that caused the problem in the first place (e.g., few remember the name of the manufacturer of the cellphone batteries that would burst into flames).

Not every company must maintain a high level of traceability, and the level of traceability is frequently directly related to the perceived liability that a company has with its products, although it is pretty certain that the cell phone manufacturer did not perceive a significant liability related to those phone batteries bursting into flames. In many cases, the full extent of liability will not be understood until an issue arises and must be challenged.

Traceability, however, is not only about liability for warranty concerns. For many companies, traceability is a regulatory requirement. These companies will adhere to *current good manufacturing practices* (cGMP) and, like many nonregulatory industries, will also use traceability as the primary tool in analysis for CI, process management, and quality control.

5.5.1 Traceability for Process Improvement

In some situations, companies rely on historians (i.e., databases that are used in supervisory control and data acquisition—SCADA, and programmable logic controller—PLC, systems) to maintain the traceability within the process. Historians store large amounts of data from sensors and actuators to record the state of equipment and tools during the time a production unit is processed through an operation. This data is related to timestamps, and if the historian is set up properly and the system time from multiple PLC systems is synchronized to the second, the data can be used to highlight particular states at the right time (e.g., the temperature of an oven when a production unit was placed inside for curing). The issue is that *time* (or, more particularly, a time range) is not the easiest data element to use as a record key when looking for correlations between product and equipment over multiple runs of production or over multiple units of the same order. It also makes it difficult to relate the data from one operation for a production unit that may have been affected through its production run in a different operation.

When data collection is set up properly, the MES can record the following process-significant data related to any production unit/operation combination:

- Product serial number (i.e., the production unit's reference ID)
- Operational start/stop times (i.e., to derive the cycle times)

- Genealogy information at each operation (i.e., the ID for all material used and when it was used)
- Production operators (i.e., who performed the work)
- Tool/equipment ID (i.e., the workstation and tools used, including multiple performance characteristics of the equipment at the time the production unit is being processed)

The issue with the term *process-significant data* is determining which of the data elements are *process-significant*. In relation to that data, the significance of an element will depend on what is being investigated. In many implementations of an MES, there is a strong desire from the implementation team to *stabilize the process* (i.e., establish all process parameters) before defining the data that will be collected and, therefore, the interfaces that will be required (human or machine interfaces). In addition to the interface, there is the issue of how to hold onto the data once it is collected. In the early days of MES, many implementations were set up to collect all the raw data that was possible "just in case." This resulted in gigabytes of raw data that were collected and never used because getting into the data was cumbersome and required dedicated Structured Query Language (SQL) coders. Even if the coders accessed the data, they still required guidance on which data elements to interpret as significant. The other problem was that, in many cases, the raw data then had to be *conditioned* (i.e., summarized or aggregated in some way) to be meaningful, and the aspect of conditioning that data was also related to what was being investigated.

Fortunately, the technology for processing data in real time has come a long way in a short amount of time, and there has been significant improvement in the area of conditioning data to make it meaningful and easier to interpret. There has also been considerable improvement in the technology to access that data in a user-friendly manner without having to write SQL scripts. This makes it much easier for manufacturing and quality engineers to access and review data to identify meaningful correlations.

With well-designed access to data and the implementation of good data collection methods, it is not necessary to collect all data all the time. The key point of designing the data collection structure for general collection purposes is to define a data model for equipment and tools that reveals changes to a broad range of data elements. The function of the general data collection model is to provide an indicator of an issue that may be related to more than one data source. The purpose of data collection in this regard is not to be able to pinpoint a problem but to indicate where to look for more information on the problem. Once the indicator is provided, it is necessary to look deeper into the raw data that is feeding the indicator. The other aspect of indicator data collection is to provide visibility to trends in the data or to determine a correlation between the data from multiple sources and events that occur on the production floor. Being able to identify correlations in data will enable Quality or Manufacturing Engineering to define a production scenario that can be tested with more detailed and targeted data collection definitions. It is very important that Quality and Manufacturing Engineering become aware of the structure of the data that is being provided by the MES.

The following sections examine the aspect of quality control for product and material control. The use of traceability for process and equipment management will be covered in more detail in Chapter 6. Supplier management and warranty management will be covered later in this chapter as well.

5.5.2 Traceability and Quality Control

Quality control (and specifically traceability) for the scope of product and material management is concerned with two situations: (1) ensuring that only the correct material went into the product and (2) ensuring the correct process was followed for the product.

The use of material IDs in one form or another has been part of material management for decades. From the production planning side, *material IDs* are frequently used to help define the set of materials (usually identified with a batch ID or set of serial numbers) to be consumed for any specific production order. Before MES, once the material was on the production floor, all aspects of control were maintained by first-in-first-out feeders or kitting as part of the LSS methodology. Any deviation in the manual process would invoke a temporary loss of control that had to be detected and corrected by the Quality control group through various inspection methods.

Earlier in this chapter, we covered product nonconformance and repair. When the system is provided with the material IDs required, an MES could alert the production operator in real time if the current LSS material supply is not correct for the production run. The operator would not be permitted to complete the operation, which results in preventing the nonconformance (and the need for repair). This is a relatively simple view of the use of an MES for traceability. However, the use of material IDs in quality control expanded to a wider capability when the MES was implemented. Additional focus on implementing more complex material ID strategies using an MES also helped improve the effectiveness of traceability and in using material IDs as a fundamental part of quality control and continuous improvement.

What came out of that scenario was the recognition that if Quality staff discovered a problem (characterized a pattern of occurrence) with a batch of material (e.g., a sudden increase in failure related to a specific material ID), then the system acting in real time could alert Quality and Production staff of the occurrence. The system (knowing where all related WIP material was located) could then be used to automatically quarantine the entire batch of material and highlight other WIP products that were made with the same material. This ability to trigger an alert based on the pattern of nonconformance (or another pattern for detection) was not recognized until recently as a major part of most MES. Further extension of this capability would be to send a signal to the enterprise resource planning (ERP) system notifying it of the occurrence so the ERP could further quarantine inventory from the same batch of material remaining in the warehouse. In the *just-in-time* environment of many manufacturing industries, this rapid detection and material control capability has also been extended to the entire supply chain.

In some industries (e.g., aerospace and pharmaceuticals), the traceability requirement is so important that many countries require, by law, traceability of all materials used during the manufacturing process. MES have proven to be so valuable in these industries that many MES capabilities have been preapproved by regulators for use within these production environments.

5.5.3 Supplier Management

Part of the extension into the supply chain and of recognizing and detecting nonconforming material is the ability to quantify the performance of the material supplier (whether it be the manufacturer of the material or the distributor). As long as the product manufacturer maintains close control of the movement of material within its production plants, it becomes easier to identify which suppliers are providing material that is consistently performing well. As consistency is a highly valued characteristic of a supplier, using MES-level information should become a major part of a supplier management program. Some suppliers will look to customers who use an MES as a significant part of their own CI program. A manufacturing company that uses an MES can provide accurate information on usage and handling conditions as part of its feedback on material performance capability.

5.5.4 Post-Manufacturing: Warranty and Customer Service

This section explores the use of MES genealogy data as part of a company's warranty and customer service programs. If a product is recalled, a company has two ways of handling it; the response depends on the detail of the information that the company maintains. In a scenario of a problem with a specific batch of material, without the use of an MES, the best a company can do is determine when that batch of material was moved to the production floor. In many situations (as a result of convenience to the materials management team), a manufacturing floor will allow the production staff to mix the suppliers and batch IDs for the material consumed during a production run. This is frequently implemented with the use of *alternate components* when loading the LSS bins with material. In these scenarios, there is little attempt to control the batch ID (or component serial number) that is consumed into the product. As a result, in the event of a recall, the manufacturing company only has a general idea of when the material was moved to production and cannot be certain as to when that material was exhausted. This drives the manufacturing company to launch an "investigation" into which products could be a problem and, at best, provide a time frame for products that "may" contain the problem material. The manufacturing company must now absorb the costs of determining the extent of the problem, inspecting products that may be involved, and repairing (i.e., the replacement of the problem material) the products. However, because there is no record of where these products have been sold, it is difficult to plan for the labor and the distribution of material to fix the problem, which adds complexity and cost to the recall process.

When an MES is used (and configured well), the company is able to accurately control the movement and consumption of material and track the genealogy of each product that is made. If a product is recalled, the manufacturing company could extract the history of the products made with a specific material batch ID, access their warranty records (presuming customers have completed their warranty registration), and create a cross-reference to the customers that would be directly affected. Using the cross-reference list, the company could then send a letter (or better still, an email) to each customer asking them to arrange for a replacement of the problem material (or entire product). Knowing which products (by serial number) have the problem material and the customers who acquired those products can greatly reduce the cost of warranty management and recalls.

Few products (or product types) escape being affected by a warranty or recall. However, with the amount of outsourcing occurring at some levels of manufacturing, the number of affected products is constantly increasing. At one time, the idea of recalling a cell phone because of a battery that catches on fire would have been a far-fetched scenario to plan for.

5.6 Additional Notes

Some of the concepts presented in this chapter are quite advanced. However, they are well within the capabilities of higher-end MES. The primary concept of this chapter is that using an MES to monitor and control the quality parameters of process and material opens the opportunity to identify, isolate, and resolve issues that affect production in real time. It is important, however, to ensure that the MES configuration has been designed with the support of these issues in mind. A poorly configured MES can be more detrimental than no MES.

The key function that an MES provides is the ability to correlate seemingly disparate events during production. The correlation of data provides visibility to manufacturing issues (that were often not known) and gives Manufacturing and Quality Engineering the guidance needed to take their production capabilities to the next level. No MES (or other enterprise system for that matter) will likely save money and provide a return on investment just by being implemented. It is the intelligent use of the MES and the data (and the proper correlation of data to information) it provides that drives improvement in manufacturing.

As a final note, it is sometimes difficult to recognize the complete definition of processes and data models that create an effective capability in quality management for a manufacturing environment. As a guideline to help companies move to a more effective management capability, the ISA-95 standard has defined a full model of activities and data structures that are needed within quality execution management for manufacturing and manufacturing operations management in general. As many of the better MES have also been designed to support these processes, using the ISA-95 standard as a fundamental model for process management can also help to ensure a more effective MES implementation and a greater level of support from an MES to manufacturing quality management.

5.7 Quality and Traceability Benefits

A summary of the benefits of MES for quality and traceability is as follows:

- Provides guidance for various QA functions
    - Calibration management
    - User certification
    - SPC monitoring
    - A platform for implementing poka-yoke methods
- Supports test and inspection tracking

- Supports legal and/or customer requirements for traceability and genealogy information
- Provides support to production repair operators
- Provides a correlation of data to support quality initiatives
- Provides accurate data for warranty and customer support

6 Production Monitoring: Details and Benefits

6.1 Introduction to Production Monitoring

In the previous chapters, we covered how a manufacturing execution system (MES) brings value to order execution, quality assurance, and material management activities and processes. All these business processes share in the benefit provided by the ability to monitor, annunciate, and trend actual manufacturing activities and events that can directly affect the efficiency of the manufacturing process. Especially in a just-in-time manufacturing environment, any amount of time that production is scheduled and cannot be met becomes an issue, and risks must be minimized. In this manufacturing environment, everything that interrupts the production schedule must be annunciated, acknowledged, and corrected in the shortest amount of time possible. Process-critical parameters, like the temperature of an oven drifting out of the range of product requirements or manufacturing equipment being down due to an unscheduled repair, can have a major impact on the on-time delivery of an order and the *cost-effective* delivery of that order. Without the capability to respond immediately, the whole manufacturing process will not be able to meet the demands of the production schedule. An MES provides these monitoring functions in support of plant operations.

An MES brings efficiencies to the production monitoring process because of its ability to detect and annunciate faults and other concerns in real time over a large geographical area as well as some remote areas. This helps speed up resolution responsiveness, provides deeper details into these concerns (and the impact of the concerns on products), and provides real-time analytics and trending functions. Without these production management functions provided by an MES, direct management of production is based on experienced guesswork or isolated pockets of information without the holistic visibility of the impact on the entire production process.

Production monitoring includes the following areas, which we will explore in this chapter:

- Equipment warnings and faults
- Production faults
- Quality faults
- Throughput and cycle time monitoring
- Overall equipment effectiveness (OEE)
- Andon
- Trending
- Process failure mode and effects analysis (PFMEA)
- Six Sigma
- Preventive/predictive maintenance (advanced)

6.2 Production Monitoring before MES

Before MES, there was production monitoring, but the annunciation of different events was limited to physical devices. Some smaller manufacturers still use this process because it provides a cost-effective annunciation in a small environment. As shown in Figure 6-1, these devices range from a stack light that signals a green (bottom), yellow

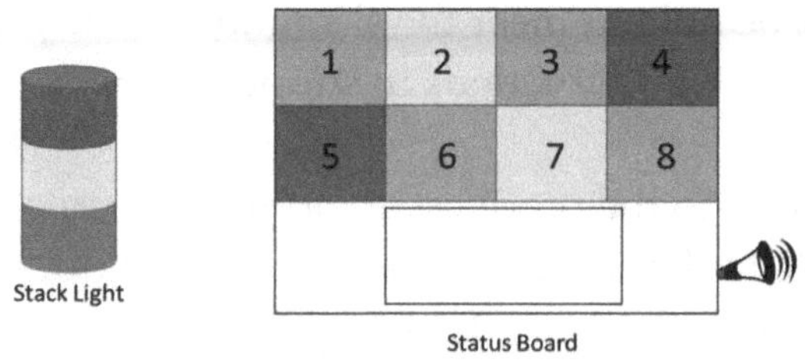

Figure 6-1. Device-based production event annunciations.

(middle), or red (top) status to an overhead board that signals specific workstation operations (green for 1, 3, 6, and 8; yellow for 2 and 7; and red for 4 and 5). Some of these devices even play a specific musical tune to alert the appropriate team to respond to the manufacturing event.

At first, these devices were physically wired to the programmable logic controller (PLC) running the operation, whether it was a tooling or conveyance system. While this solution works in small manufacturing plants to respond to events quickly, it lacks any analytics, historical functionality (once the event is done, the data cannot be recalled), or trending functionality. It is also limited to the actual physical placement of the annunciation hardware and cannot leverage any functionality that would provide remote notifications of events. In addition, it can only signal that there was a type of fault; it cannot provide real details about the fault. Additional research would be required to determine the root cause of the fault.

6.3 Monitoring Area Details

Before we get into the different areas within production monitoring, it is useful to discuss planning production monitoring and the level of detail required. This planning has a direct effect on how an MES will be configured and how it will operate. First, when it comes to monitoring, planning, and determining the equipment failure modes, the equipment vendor usually lists the faults first because they must know the operational details of their equipment. Additional equipment faults can be added through planning activities between the plant and the vendor, especially for faults in which the plant-floor equipment loses communication with the higher-level MES. Some equipment, especially conveyance systems, has programming that supports production metrics and faults, which can then be easily added to the MES configuration. Once the failure modes are determined, it is up to the plant to determine how the MES will be configured to annunciate to the proper response team and what preliminary action must be taken. The annunciation solutions available depend on the defined process of the response team and the system's functionality. Solutions can include local PC clients, smart device texting, smart device applications, and emails.

Planning for equipment monitoring also includes the ability to use the data to provide useful analytics and trending functions. While these functions are not necessarily related to faults, they can provide significant insight by delivering the data through organized displays like dashboards and reports or through similar distribution methods. Delivering data this way supports continuous improvement and provides visibility of long-term trends in production events.

As part of the interfacing strategy for monitoring production resources and equipment, Parts 4 through 6 of the ISA-95 Series define the messaging structure and response to any faults that may occur on the production floor. The following section provides an overview of the types of faults and examples of how they can be referenced in an MES.

6.3.1 Equipment Faults

The first and foremost concern in production monitoring is the ability to know the health of all production equipment in real time and to correct faults in such a way as to minimize production downtime. Every minute that production is down due to an equipment fault is time lost in the fulfillment of the production schedule. This includes every piece of equipment that helps produce the end product, including all tooling, conveyance systems (horizontal and vertical lifting systems), robots, PLCs, operator-assisting equipment like lift assists, and any other piece of equipment that directly supports production. Equipment health is based on the detail of the faults of that sub-system. The more detailed the fault notifications are, the easier it is to determine the root cause of the fault and correct the fault quickly.

Equipment faults are classified into three main prioritized categories:

1. **Production critical** – Production cannot continue until the root cause is fixed. This requires immediate attention and resolution (e.g., a motor drive overload that has stopped a conveyor or tooling with an electrical welder overcurrent).

2. **Production major** – Production can continue only because there is a defined workaround. Production throughput and data collection are usually negatively affected. While this does not require immediate attention, it does require quick attention and root cause resolution (e.g., a nonoperational conveyor position sensor or a tooling's operational temperature drifting outside the normal zone, neither of which has been confirmed to not cause a quality issue for this instance).

3. **Production minor** – Production can continue, but at some point in the future, the root cause will become either major or critical (e.g., a conveyor drive motor bearing over temperature or a tooling temperature sensor requiring a calibration check).

It is essential to have a dedicated shop floor maintenance team that is responsible for responding to equipment faults. Therefore, all annunciations to the team (depending on how the business process is defined) must be determined as part of the available faults.

System-based faults can also be considered as part of equipment faults. These faults can range from simple issues, such as production data (or order information) not being available, to critical level faults, like the application system being down. Typically, the shop floor maintenance team does not respond to these faults; instead, the information technology (IT) support team responds. Similarly, the business process will determine how these faults are annunciated and may also involve the customer when order data is not delivered properly from them.

As mentioned before, *equipment faults* are typically defined first by the equipment vendor and then enhanced by the manufacturer's requirements. It remains the sole responsibility of the manufacturer, not the vendor, to determine how best to annunciate the events in real time.

One more benefit that an MES can provide to equipment fault monitoring is the ability to interface with maintenance management software tools (which are frequently becoming part of ERP) to facilitate managing the plant maintenance business functions. The data that pertains to direct maintenance intervention in resolving equipment faults includes time to resolution, parametric data for root cause analysis, and other related data elements, all of which can be sent to the maintenance management software. When coupled with the existing scheduled maintenance, this data gives a clearer picture of the maintenance activities to provide. Another function that MES equipment faults help to provide is the end link to the replacement of faulted devices that support production. Maintenance must keep an inventory of any device that is critical for production, and knowing the history of equipment faults will help keep this inventory properly sized. As mentioned in Chapter 4, all inventory must be carefully managed because inventory increases cost and requires real estate in the plant for storage; more inventory incurs more costs and storage space. These costs are not value-added to production.

6.3.2 Production Faults

Production faults slow down or stop production and are not caused by equipment failure. These faults can be classified into two categories: (1) operator-based faults and (2) buffer-zone-based faults. Operator-based faults include work halted due to the following:

1. Parts not being available in station line-side bins
2. Operation overcycle (Production takes longer than the defined cycle time.)
3. Starved work-in-process (WIP) delivery (The previous station has not sent the product.)
4. Blocked delivery to the next station (The next station has not been cleared.)

The last two delivery faults can also occur if a physical buffer is available. When this is true, the buffer-based faults are as follows:

1. Starved WIP delivery (The previous station has not sent product; the buffer is empty.)
2. Blocked delivery to the next station (The next station has not been cleared, causing the buffer to fill completely.)

Buffer zones are used when one production area has a different production rate than another (see Figure 6-2). Simply put, a buffer zone is an area in which subassemblies are stored temporarily; the number of subassemblies is monitored to ensure the buffer zone does not starve the next operation or block the previous operation.

In the example shown in Figure 6-2, the maximum number of parts in the buffer is 10, and depending on the configuration of the monitoring system, the buffer will enter into a fault state if the number of parts in the buffer is larger than 8 or less than 2. In this scenario, the faults could even be divided into *warnings* and *faults* to account for when the buffer levels go beyond predefined limits (a warning) and the buffer changes to being either full or empty (a fault).

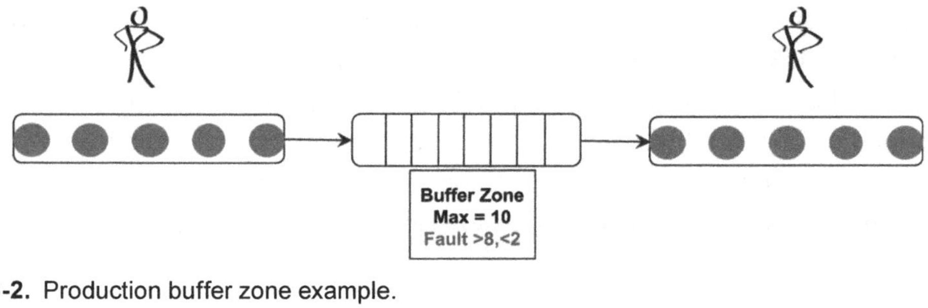

Figure 6-2. Production buffer zone example.

In all cases, responses to these production faults are usually not performed by the plant or IT support teams but rather by the production team responsible for the specific business process area. Examples include line-side part bin replenishment issues that require the plant's materials management team to respond, issues with the production schedule within the MES, or issues with an improper order launch that require the scheduling team to respond. The Manufacturing Engineering team usually responds to production overcycle events. However, there can be several possible causes for overcycle that require multiple teams and a planned response. This response may also be driven by production data analytics and trending to determine ongoing or chronic issues. We will cover this area in more depth in Section 6.3.4,

"Cycle Time/Throughput." These are just some examples; there are others we cannot fully document here.

Depending on the type of equipment provided by the equipment vendor, the logic for some production faults may already have been integrated into the equipment controls, especially in areas where a production fault, like the proper product placement in a placement fixture, is incorrect and requires an immediate response. The plant's Manufacturing Engineering team must implement additional production faults to ensure that all events are annunciated properly and therefore responded to in real time. This depends on the responsibilities of the maintenance team versus the production support team.

6.3.3 Quality Faults

Although we discussed quality faults in Chapter 5, we must cover them in additional detail as they pertain to production monitoring. This detail will define how an MES is configured and how it operates. In many cases, *quality faults* are considered a subset of production faults because capturing and resolving them usually means stopping production and resolving the faults in the workstation where they occur. This is because it is quicker and cheaper to resolve quality issues that way. However, because the Quality team is usually the team that responds to quality faults, and because quality faults are directly product-related, unlike the other production faults, we will cover these separately.

In normal operations, quality faults, as they pertain to production, fall into three categories:

1. Quality failure due to process error
2. Quality failure due to an assembly error
3. Quality failure due to material performance

Process quality faults are faults in which the manufacturing process was not followed correctly or not fully defined. An example is a process in which the product is heated to a specific temperature. If the *running* temperature is not within the specified tolerance, then the quality of the product will be affected. Examples of quality faults due to assembly include the selection of an incorrect subpart to be installed on the end product or the use of an incorrect batch of material during production. Quality failures detected by testing are not issues that the production operator would expect to see as part of the operation; rather, they represent major quality issues where the product must be scrapped or dismantled and rebuilt later in the process. This also applies to quality failures identified during inspection. Because testing and inspection occur separately from the workstations where work is conducted, testing and inspection faults—although they can be caused by errors in process or assembly—are frequently a result of material performance and are typically reported separately. The response here may require not only the Quality team but also the product engineering team, the production support team, and even the equipment maintenance team to come together and resolve the issues. We will cover test and inspection faults in more depth in Chapter 7.

These test and inspection faults are usually defined by the product engineers (as the product developers and experts) and by the quality experts, all of whom understand and help ensure the correct manufacturing process is defined. These faults will be annunciated to the proper team as defined by management.

6.3.4 Cycle Time/Throughput

Another consideration for monitoring production health is throughput and how it is affected by cycle times. This data is used to support investigating production overcycle events. An MES provides value here by collecting the data and providing the analytics to support *issue recognition and resolution.* Although this topic was covered in Chapter 3, we will review it here as it is also a function of production monitoring. *Cycle time* metrics are determined by comparing the operational cycle's actual time to its target time. *Throughput* is the measure of the rate of production in a given period of time. Closely related to cycle time is *production throughput.* Typically, throughput is measured per hour, per shift, and per day. The throughput of a line will be directly related to the operation that has the longest cycle time. In the Theory of Constraints, this is referred to as the *constraining operation* or just the *constraint.* While these metrics usually will not trigger a production fault that requires immediate attention, they will indicate when production may be slowing down enough to negatively affect the ability to meet the production schedule. These types of metrics are frequently presented on billboard-type displays called *takt boards* and can be the first indicators to production that there is a concern.

To explain this further, we must explore a real-world example. Using the buffer zone in Figure 6-2, we see five stations in a production line. It is determined that in order to support the production schedule, production must produce 60 units of a product per hour (also known *as jobs per hour—JPH*), which means one unit is coming off the line every minute. This means that the product can stay in each station for only 60 seconds, and this time must include time to complete the work required in each station and any reset or setup time to prepare for the next cycle. Given this set of requirements, Table 6-1 is created for a specific cycle.

Table 6-1. Cycle times example.

Station	Work Time	Prep Time	Total Time	Actual Work Time	Actual Prep Time	Actual Totals
1	45	15	60	45	14	59
2	40	10	50	40	10	50
3	55	0	55	56	0	56
4	50	10	60	55	10	65
5	45	10	55	45	10	55

As shown in the example in Table 6-1, for each cycle, the actual times are monitored and trended. Cycle times at or under the expectation will show up green. Although 5 seconds overcycle for Station 4, which would show up red, does not sound like an issue, if this actual cycle time repeats, it represents a production issue. If Station 4 is 5 seconds overcycle for each cycle, after 12 jobs the throughput will drop by 1 job ($5 \cdot 12 = 60$). This will cause an inability to support the production schedule. Such an event will require additional research because the root cause may be equipment-related or operator-related. Perhaps a new operator has taken over that station and is taking 5 additional seconds to complete the tasks while ramping up to speed.

Note that in this example, the overall throughput is determined by the slowest station (as stated earlier, called the *constraint*) because it is a serial production line. This demonstrates the importance of ensuring that the workload is properly distributed or balanced over the workstations as evenly as possible (*called level loading*) in certain manufacturing environments. In other environments (e.g., job shop production plants), the production floor is frequently divided into work centers (multiple workstations that perform the same work), and it is not possible to maintain the serial production flow. In these environments, throughput can only be measured at the production floor level (not the line level), and the independent cycle time of each work center must be monitored.

How cycle time and throughput are measured will strongly depend on the structure of the production environment; the MES application will likely be designed to support one environment or another, and reporting on cycle time and throughput must be configured specifically for that environment. Without an MES, collecting and organizing data to determine where issues may occur becomes a manual, labor-intensive process that is typically performed in a spreadsheet.

6.3.5 Overall Equipment Effectiveness

OEE is a metric that provides a measurement of the effective use of a resource and is the combination of equipment uptime, production speed, and first-pass quality compared to expectations from the equipment's scheduled time. The data required to calculate OEE is extensive, and if collected manually, it is prone to errors and interpretations that may make the metric of little value. An MES can calculate this metric easily based on its own internal data tracking and calculations. OEE is calculated in the following equation:

$$\text{OEE} = (\text{equipment availability}) \bullet (\text{production performance}) \bullet (\text{first-time quality rate})$$

where

$$\text{equipment availability} = \frac{(\text{scheduled production time}) - (\text{equipment down time})}{\text{scheduled production time}}$$

$$\text{production performance} = \frac{\text{actual jobs per hour}}{\text{target jobs per hour}}$$

$$\text{first-pass quality rate} = \frac{\text{actual good product}}{\text{total product produced}}$$

Jobs can be given per shift, per day, and so on, provided the timeline is matched throughout the formula. Good product refers to product that is not scrapped or does not need to go into repair or rework. In other words, products that are acceptable from performing each operation only once, with no quality dispositions. Figure 6-3 shows an example of calculating OEE per hour.

In Figure 6-3, equipment availability is determined by subtracting the unscheduled downtime from the scheduled available time. The scheduled JPH is determined by the maximum line speed (60 JPH) times the expected available time (minus any scheduled production downtime like breaks and lunch). The actual quality is the count of end products that completed a first-pass quality check, where there were no issues after the initial manufacture that would have caused repair or rework activities. Due to the number of calculations involved, an MES provides value by performing the calculations and distributing dashboards and reports. As OEE is an operational performance metric, it is important to calculate it in *near* real time and therefore calculate it from the operational database (as opposed to the data warehouse) as described in Chapter 9.

OEE Calculation per Hour Shift = 8 hours									
Scheduled JPH	60								
Hour	Downtime	Availability	Scheduled JPH	Actual JPH	Performance	Actual Quality	Quality Rate	OEE	Reasons
1	15	0.750	60	40	0.667	39	0.975	0.488	Equipment Down for Repair
2	0	1.000	60	61	1.017	58	0.951	0.967	
3	0	1.000	45	46	1.022	45	0.978	1.000	15-minute break
4	0	1.000	60	62	1.033	62	1.000	1.033	
5	0	1.000	30	30	1.000	29	0.967	0.967	30-minute lunch
6	0	1.000	60	55	0.917	55	1.000	0.917	
7	0	1.000	45	42	0.933	41	0.976	0.911	15-minute break
8	0	1.000	60	58	0.967	57	0.983	0.950	
Shift Total:	15	0.969	420	394	0.938	386	0.980	0.890	

Figure 6-3. OEE calculation per hour example.

6.3.6 Andon

Andon is a term first coined by the Japanese that refers to a systematic approach to annunciating production and equipment events. We are covering this topic here because, depending on the configuration, Andons can be an annunciation tool for *all* production events. In the original concept, the Andon board was a physical, lighted board hanging over the production area and controlled by the physical input/output of the PLC. It was used to display the state of the workstations in a manner similar to that in the status board shown in Figure 6-1.

As shown in Figure 6-4, the objective of Andon is to create a near real-time dashboard to show key production metrics such as cycle time, cycle time overage, and OEE (similar to the status board shown in Figure 6-1).

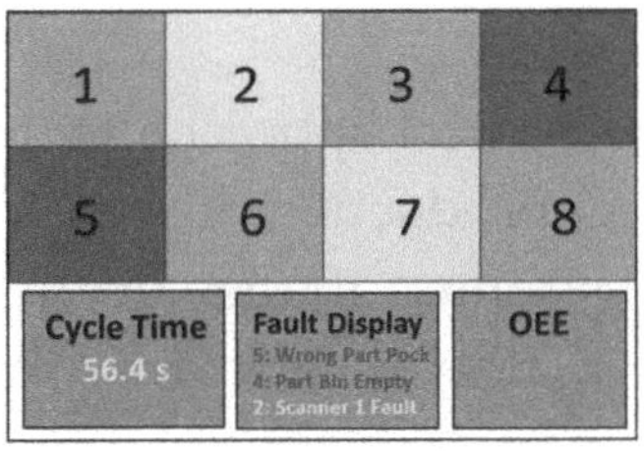

Andon Dashboard

Figure 6-4. Sample Andon dashboard.

Other information points that can be added to or changed on the dashboard under Manufacturing Engineering discretion are first-time quality rate (also called *first-pass yield*), actual JPH versus schedule JPH, and a display of the current prioritized faults (or alarms, depending on plant preferences). This Andon screen can be displayed on any monitor in the plant, including monitors that physically hang over the production areas like the status annunciation screens described earlier. Most MES have templates that can be used as the basis of Andon functionality. Care must be taken to ensure that (1) the data needed by the Andon is properly configured to be available in the MES and (2) the available data is properly interpreted by both Operations and the MES implementation team.

6.3.7 Trending

With the ability to send key production data to a database (that can include cycle times, fault events and durations, and other key process-sensitive data such as production temperature readings), it is possible to create intelligent reports configured to monitor and detect trends that occur during production. This is useful to help expose ongoing or chronic issues from all areas of manufacturing (i.e., equipment, production, quality, and throughput).

Examples of trend reports include the following:

- Top 10 equipment faults
- Top 10 production faults
- Cycle time graph per defined period
- Process variable graph per defined period (i.e., temperature)
- Equipment fault time to repair
- Mean time to repair

6.3.8 PFMEA

PFMEA refers to a planned, methodical approach to anticipating and managing any event that could negatively affect production. Its purpose is to determine and define the ways a process may fail (the failure mode), detail the severity and describe the effects of the failure mode, and identify the corrective or preventive action required.

Failure modes can range from full failure to degradation of service. Their severity can range from the loss of primary functions that causes stoppage to annoyance, and the failure modes may also include regulatory and/or safety implications.

Once the PFMEA is defined and used to implement which faults are generated and monitored, MES production monitoring can be used to track the actual PFMEA issue, and Manufacturing Engineering can use MES trend reports to analyze PFMEA activities.

A small part of a PFMEA is shown in Table 6-2. More details on PFMEA and MES will be provided in Chapter 7.

Table 6-2. Mode: loss of conveyor function.

Failure Mode	Severity	Reason	Fix	Effort (min)
Conveyor Line Stops	High (Line Stoppage)	Motor overload	Replace fuses	10
			Check and/or replace wires	10 or 45
			Replace motor	30
		Variable frequency drive fault	Reload last good configuration	10
			Replace drive	30

6.3.9 Six Sigma

Six Sigma, first created in 1986 and implemented by General Electric in 1995, is a quality improvement methodology that seeks to improve the output quality of a process by identifying and removing the causes of defects and minimizing variability in manufacturing and business processes. Typically, it is executed by prioritizing the impact of product defects and creating projects to either eliminate or minimize the impact of the root cause of the variation. This is just one type of process improvement method.

MES production monitoring supports Six Sigma efforts by tracking the following: process/product defects, process-critical parameters and equipment faults that affect the creation of defects, and production-specific faults. Tracking these items, in conjunction with nonconformance quality tracking as detailed in Chapter 5, can provide and correlate the data needed to generate the analytics required to not only prioritize the causes of defects but also provide the data required to make a root cause determination.

Six Sigma and MES will be covered in more detail in Chapter 7.

6.3.10 Preventive/Predictive Maintenance (Advanced)

Now that we have covered the basics of production monitoring, we must cover more advanced topics and functions. Not only is it necessary to annunciate actual equipment or process fault events but it is also beneficial to annunciate maintenance activities that can be scheduled before a potential fault event occurs. This functionality is called *preventive maintenance.* For example, consider a power relay used in a system to provide switched power to a critical process that is only rated to operate for up to 100,000 power cycles. Once the cycles reach 90,000, there could be an activity scheduled to replace the relay. The benefit of the MES is that it can track the number of cycles and the annunciation of the parameter crossing the warning threshold. Another benefit example is a conveyor drive motor that is rated to operate up to 800,000 hours under full load and can be scheduled for replacement when its life has reached the 750,000-hour milestone, preventing the unexpected shutdown that could have occurred.

Another advanced production monitoring function is *predictive maintenance,* which takes the concept of preventive maintenance further. System implementers will develop additional logic that determines when the performance of a device or process is starting to decline and has a potential for failure. This logic is based on trending and historical data. The data is typically provided by the equipment vendor regarding the mean time between failures (MTBF), and the logic can be developed as part of the original implementation or after the customer has had enough experience to derive the data themselves. This logic can be used to determine, annunciate, and even schedule the appropriate maintenance activity.

As shown in Figure 6-5, the chain density is monitored as it stretches over time. In this case, history from MES records may have shown that every time the density falls below the value of 2.0 links per inch, the chain breaks. This causes downtime because the chain must be replaced. However, it is not always failing at the 300-day mark because the conveyor does not run every day. A predictive maintenance solution would be to track the density loss, and when the density loss approaches the historic

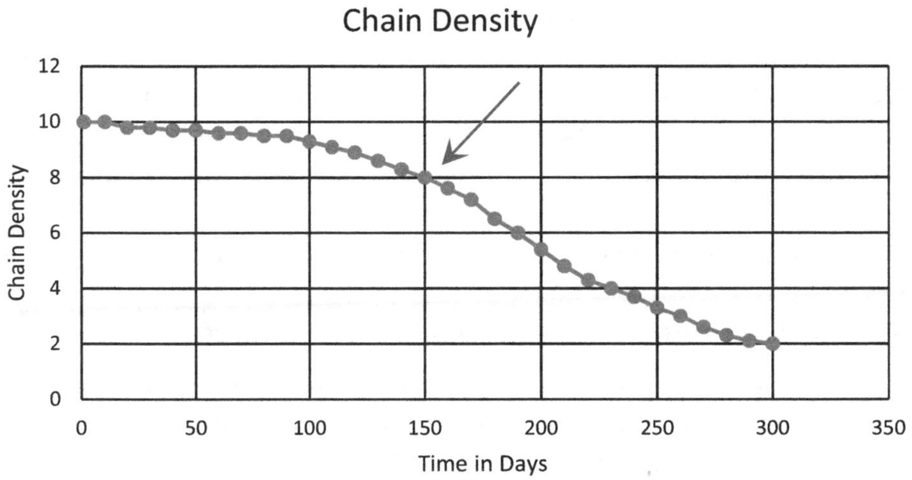

Figure 6-5. A predictive maintenance example: conveyor chain trend to failure.

value, the MTBF can be calculated and a chain replacement can be scheduled. For the longer term, replacements with a chain that has a longer stretch characteristic would be beneficial so that the chain does not need to be replaced as often.

Determining the trigger value for any predictive maintenance is a management decision based on the historic data. The data determines the likelihood of failure and the proportional increase in risk. Management policy determines how much risk to accept when management compares the cost of changing the chain to the cost of failure.

Many predictive maintenance algorithms have been determined, developed, and implemented by equipment vendors because they have reviewed their equipment's historical performance for years and have developed the algorithms to respond. Sometimes a plant, through its own monitoring and trending diligence, has been able to develop some of these solutions as well.

In both preventive and predictive maintenance examples, not only can an MES provide value by providing data collection and calculation abilities, but it can also be configured to annunciate to the proper team. Additionally, the MES can be interfaced with other maintenance scheduling systems to schedule maintenance activities automatically. This process helps to optimize the allocation of maintenance resources.

6.4 Production Monitoring Benefits

In summary, the benefits of monitoring production with an MES include the following:

- Real-time annunciation of faults (including equipment, production, and quality) which decreases the time to resolution
- Support of ERP maintenance management with data on unscheduled repairs and resource usage
- Throughput and cycle time annunciation, increasing responsiveness to potential time issues
- The ability to monitor the trend OEE metrics and other key performance indicators, increasing trending responsiveness
- The ability to display all active real-time production data in an Andon-specific format, increasing awareness and responsiveness to issues
- Historical and trending reports, giving the ability to diagnose and solve issues

- Support of PFMEA execution and improvement activities through analytical reports
- Support of Six Sigma improvement activities through analytical reports
- Support of advanced preventive and predictive maintenance functions, providing even more responsiveness to potential production downtime events before they occur

7 Continuous Improvement: Details and Benefits

7.1 Introduction to Continuous Improvement

There are those who would use the cliché that *continuous improvement* (CI) is about the journey, not the destination, and to some extent, this is true. A line of thought that goes along with the journey concept is that the level of improvement obtained by each step must always be worth the expense to achieve that step. This is where Quality and Operations management frequently collide. Those who are *Deming purists* will state that the benefit will always outweigh the input because of *nonmeasurable customer satisfaction* characteristics. From an operations management viewpoint, if you cannot measure it, how do you know you have improved it? The bottom line is that the cost of the improvement for a product/process combination *must* be recoverable within the life of the product, or it will result in a net loss for the company.

It can be argued that changes in product characteristics can improve the desirability of the product from the customer's perspective, but these issues are under the control of Product Design and Marketing. Therefore, these aspects are outside the scope of this book. From a manufacturing perspective, the only aspects of improvement controlled by operations are the efficiency and the ease with which the product can be made. The ease of making the product is the primary driver for the initiatives of design for manufacturing (DFM) and is a collaboration between Manufacturing and Product engineering. There will be more on this later in the chapter.

In the aspect of process efficiency, you must *standardize, standardize, standardize.* In the context of the value gained by any improvement, you must know your costs. In a cost/benefit analysis, as previously mentioned in this section, if the cost of the improvement is greater than the benefit, then the improvement should probably be avoided and a mitigation strategy should be used instead. The only caveat is the need for a clear understanding of the benefit to be achieved.

By now, you are probably thinking, "What does this have to do with MES?"

This chapter discusses the aspects of a good CI program and how MES can be of value to CI initiatives. Regarding the issues we just raised, this chapter will highlight some high-level concepts that will keep the CI program on track and provide input on observations we have made over the years.

7.2 Brief Introduction to MES Configuration to Support CI

One of the key factors to a successful CI program using an MES is how the virtual production environment is configured within the MES database. When the MES is first implemented, not all the data needed to manage the process or for data mining will be known. As such, it is important to configure the MES environment during the initial setup to provide the most flexible configuration possible. Failing to provide a flexible MES environment from a data configuration perspective can limit the visibility that companies have into the root cause of a problem. In this section, we give a quick example of a typical configuration concern.

In Figure 7-1, a simple process of four operations has been set up. The figure displays a common configuration error implemented by teams that are highly enterprise resource planning (ERP)-centric. In this system, the implementation team has chosen

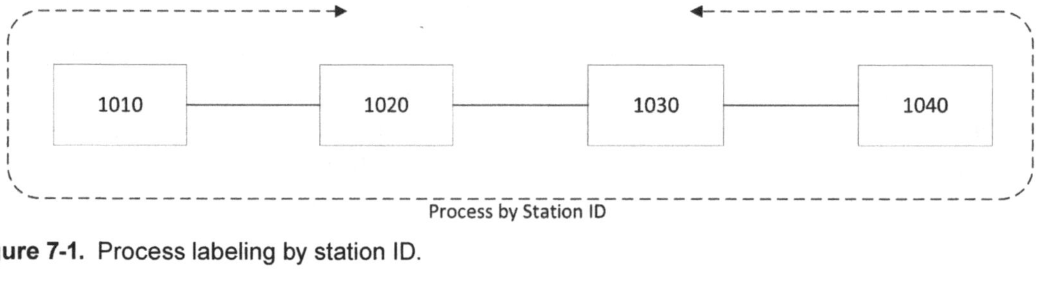

Figure 7-1. Process labeling by station ID.

to define the operations within the process according to the station identification that is performing the work (which is usually derived from the ERP configuration). It considers the operation identification and the workstation identification as a single data element. Another common implementation concept that accompanies this mindset is that the station identification is also used as the ERP reporting step ID that is set up to record consumption in the ERP system (e.g., station 1010 in the process is related to ERP reporting step 1010 in the ERP consumption routing). The problem with this configuration is that it fails to recognize that the workstation layout on the floor may change, and if another line is needed (regardless of whether or not it is in the same plant), a new routing must be defined for the configuration.

Conceptually from the perspective of ERP, the example in Figure 7-2 is acceptable because consumption from either of the workstations labeled 1020 would probably report the same data. However, the additional workstation 1020 may have different performance characteristics that Operations would need to reconcile (e.g., the additional station may have performance issues that would be hidden at the data level). Additionally, having two operations on the MES routing with the same identification would cause other concerns from a data integrity perspective.

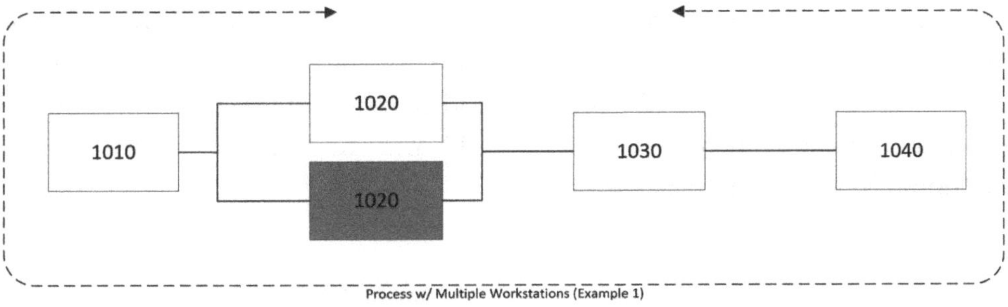

Figure 7-2. Example 1 configuration problem.

An easy fix for this problem would be simply to rename the additional workstation; however, this would require a change in the routing in both the MES and ERP and would create reporting issues within the ERP structure (see Figure 7-3). Because of the different station identifications, investigating an issue with the second step in the process, the solution would probably require the use of something called *substring searches* in the data mining, which would make the reporting extremely slow. Implementing an additional workstation in the future would complicate this issue even more.

To ensure that the configuration of the virtual production floor within MES provides the most flexibility to data mining requirements, the configuration should provide the capability to have a distinct data object for each element of the production floor.

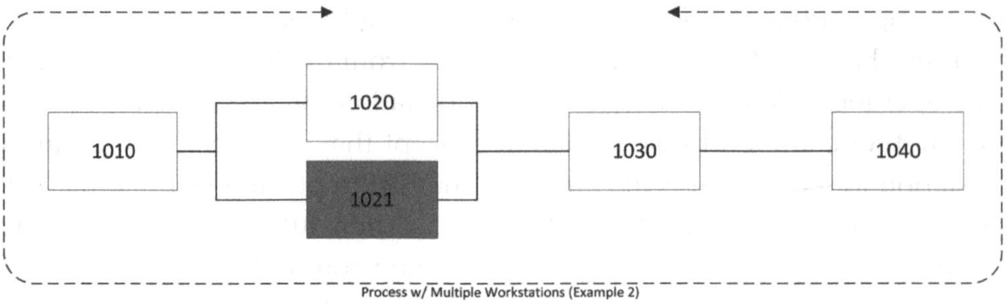

Figure 7-3. Example 2 configuration problem.

In Figure 7-4, the same process has been reconfigured to provide maximum visibility for data mining. The example operation names provide guidance as to the work that is performed at the operation (other naming formats can be used), and the workstations are independent of the operation, identifying where work is being performed. This enables operation-naming conventions that make it easy to compare the performance of similar processes. Additionally, in this configuration the workstations can be repurposed or extended as required (depending on the line or plant requirements) without changing the routing of the process. As the workstation will likely be used for other similar tasks, there will be visibility to the equipment performance at the workstation level across multiple product processes (the operation from the routing can be referenced to gather data from any workstation performing that operation). The ERP reporting step is an additional data configuration at the operation level. This enables the planning group to set up ERP consumption points only at critical locations in the process (reducing the transaction load on ERP) and modify the consumption points again without changing the process.

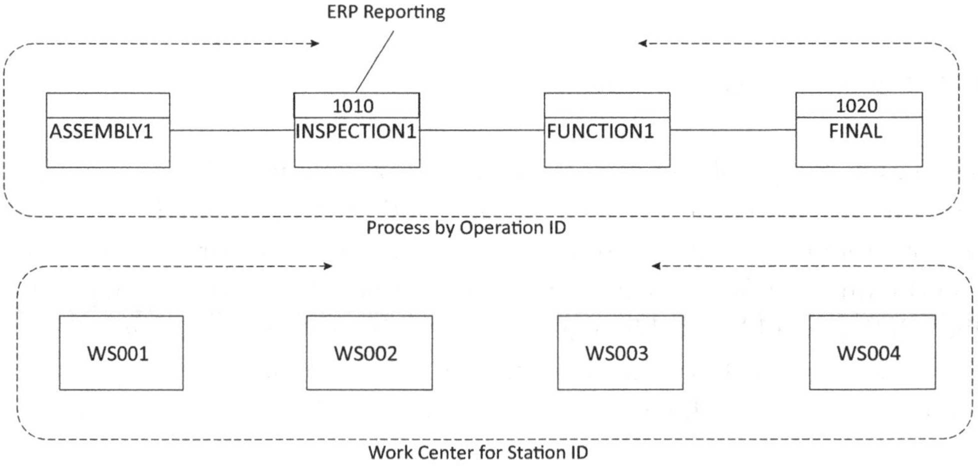

Figure 7-4. Sample MES configuration.

This topic will be discussed in greater detail in later chapters.

7.3 The CI Program

Everyone talks about it, and most people have a pretty good idea of what CI is. The issue many companies struggle with is understanding that CI is just that—an attempt to improve the product and/or the process of making the product in a continuous and methodical manner (key words being *continuous* and *methodical*). Very few companies maintain a formal CI management program, and as a result, many of them fail to achieve a sustained improvement over a long period.

A good indicator of this issue is evident when reviewing the characteristics of a product/process combination. Has there been an improvement in the cost and/or throughput year over year? If this is not the case, there are problems with the CI program.

Methodologies like Six Sigma and Lean have been developed to try to counter this issue, but the primary drivers go beyond the scope of just the methodology.

One of the drivers for this lack of consistency in managing an effective CI program is that the data required to support a CI initiative is frequently difficult to obtain while still managing production activities. In addition, as stated in Chapter 5, while a production operator is being monitored by a manufacturing or quality engineer, there is a strong tendency to *follow the process* more diligently than they would during normal production. This behavior of production operators, of course, causes the results of any investigation to be skewed and ultimately fails to achieve the expected goals.

Another driver (although less recognizable) is a lack of holistic visibility and an understanding of the process of making a product. When reviewing the process for a product, many manufacturing engineers will monitor a specific operation of the process in an effort to improve that step. Frequently it is not recognized that improving one step in the process may degrade the performance at another step.

One of the benefits of using an MES is that it records considerable data in relation to the process, and it correlates that data in the background directly to the related production events and activities. If the system has been configured to record equipment performance data automatically (as opposed to the operator providing the data), the configuration of the data recording can be updated and/or changed without any impact on the operator. In addition to background data recording, the correlated data can also be used to highlight the impact of changes to process steps that are affecting the performance of downstream operations. Providing the holistic visibility of changes to either process or product and the capability of before/after analysis are huge steps forward in managing an effective CI program.

7.3.1 The Process of Process Improvement

Most operations people are familiar with the Deming cycle of *Plan, Do, Check, Act* shown in Figure 7-5. This basic model of the continuous improvement process is a part of every quality and process improvement methodology (Lean, Six Sigma, Theory of Constraints, etc.) in one form or another and is fundamental to most analysis processes.

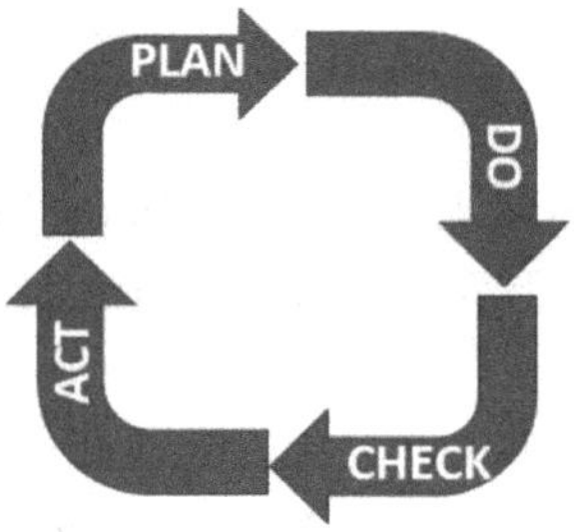

Figure 7-5. The Deming cycle of Plan, Do, Check, Act.

But is CI really that simple?

In most cases (and some would say *in all cases*), multiple loops of the Deming cycle are required to identify a solution to the problem. It will take multiple loops to determine the root cause and additional loops with a design-of-experiment (DoE) study to determine the correct solution. The number of iterations of the Deming cycle depends highly on the quality of the data and the integrity of the analysis process.

Difficulty in data collection means more Deming cycles are required to fully analyze a problem, and a lack of process integrity could result in skewed (or blatantly wrong) analysis results. Both the extended cycle process and the lack of integrity can result in it taking longer to identify the root cause of a problem and the correct solution.

As shown in Figure 7-6, depending on the problem being analyzed and the number of people involved in the initiative, the cost of iterations can quickly add up, and this does not include the cost to Operations caused by the original problem.

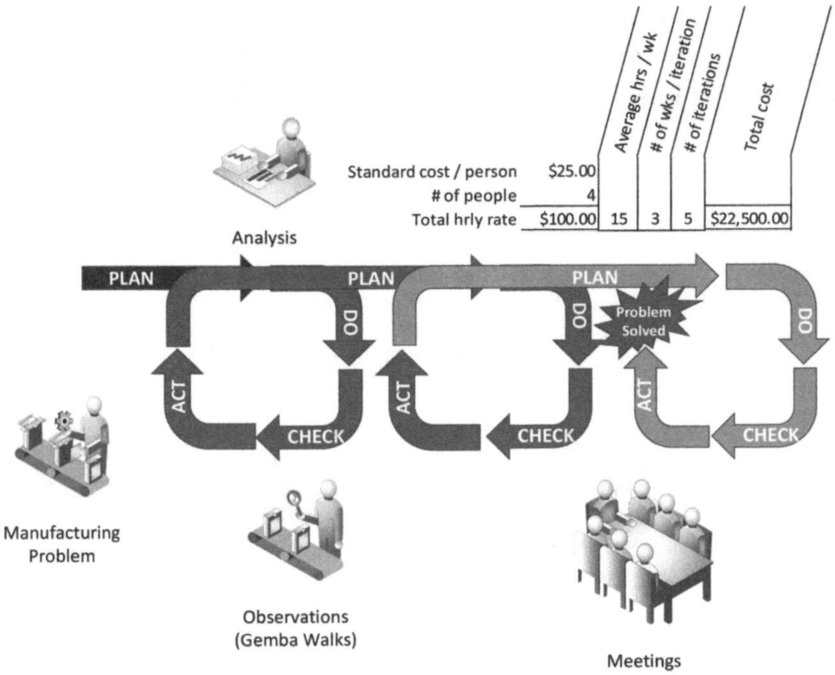

Figure 7-6. Example of the cost of multiple cycles of Plan, Do, Check, Act.

The cost of analysis gets even worse, as seen in Figure 7-7, when each department performs its own investigation of a problem, interprets the analysis, and creates a solution localized to its own scope of influence. The result is a significant amount of money invested in a limited analysis to fix a limited scope (or the incorrect scope) of the problem.

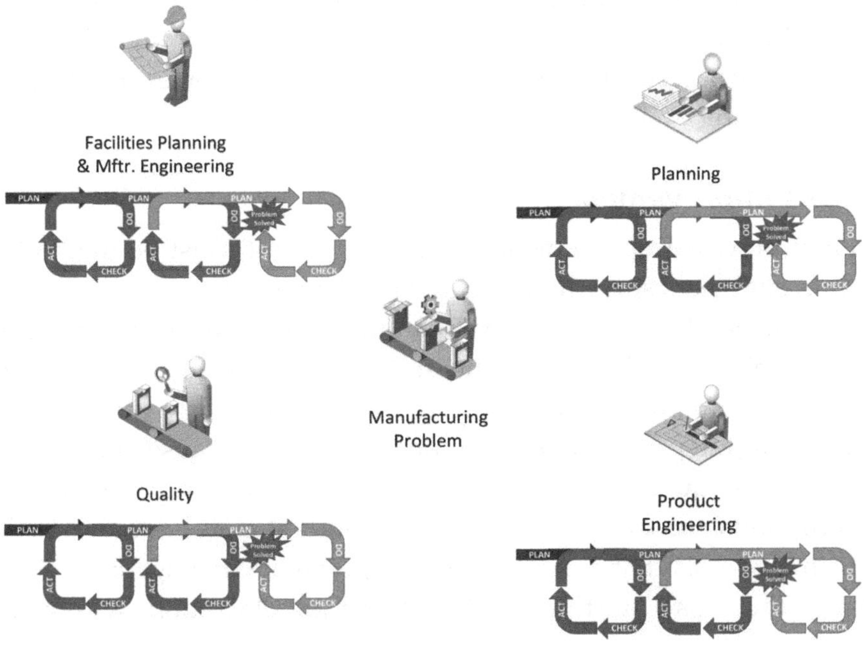

Figure 7-7. Multiple departments with Plan, Do, Check, Act.

In most cases, as shown in Figure 7-8 (and also discussed in Chapter 3), the data required by many of the analysis initiatives from different departments is the same fundamental data that has been analyzed and reported in a different manner.

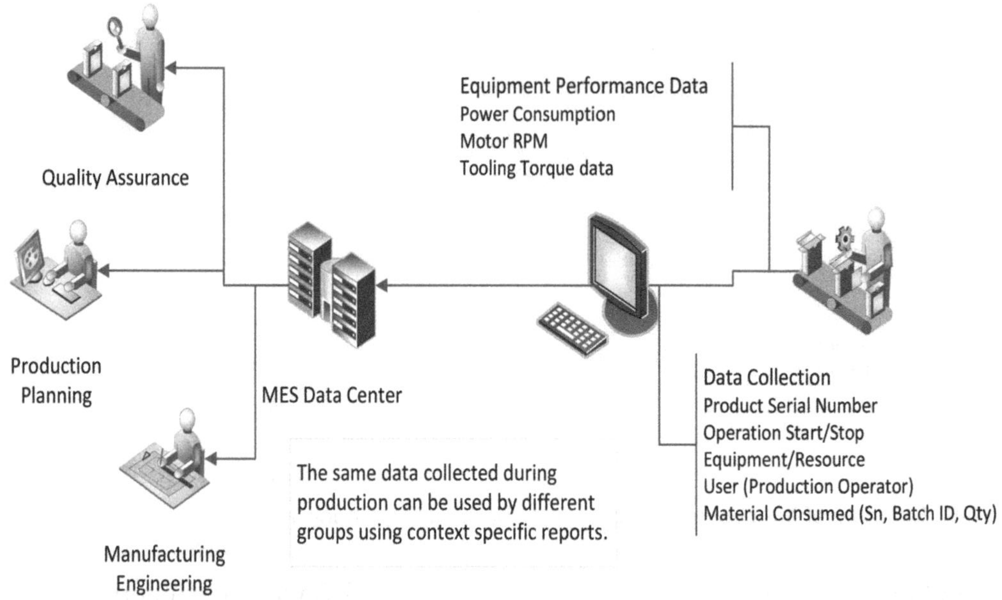

Figure 7-8. Data collection by an MES for Plan, Do, Check, Act.

Using an MES and a well-defined data collection method, the data can be formatted and analyzed to meet the requirements of all departments. In addition, the reports provided by an MES can ensure that each department is interpreting the data correctly and holistically. The end result, as shown in Figure 7-9, is a lower cost and more effective use of the Deming cycle.

7.3.2 Process Improvement

The key to any process improvement initiative is that it must be methodical. Many manufacturing engineers state the characteristics of a process based on assumptions. A common theme of all process improvement methodologies is to challenge assumptions through empirical evidence. Whether it is Lean, Six Sigma, Theory of Constraints (TOC), or any other quality methodology, a fundamental part of the methodology is to experiment with the process to determine both the root cause and the appropriate solution. Each of the quality methodologies views process improvement differently and has its pros and cons. Any (or all) of the methods should be recognized and used as one of the many tools that are available to Operations management. Selecting any one of these methods as a starting point will provide Operations with a structure to

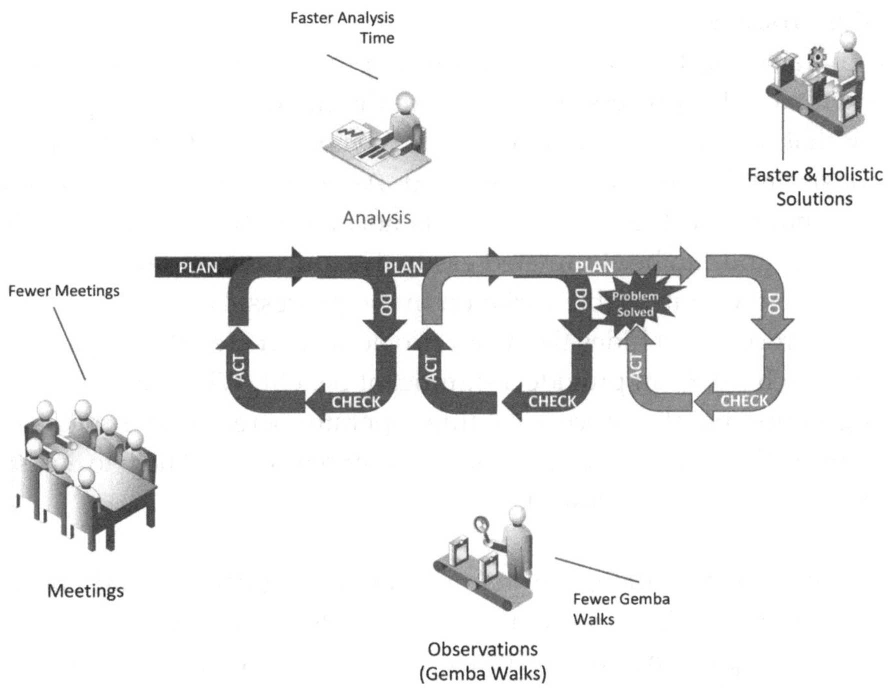

Figure 7-9. Holistic Plan, Do, Check, Act.

methodically improve. However, Operations should not rely on any one of these methods to define *the golden rules.* Operations must understand the strengths and weaknesses of each method and learn to supplement the weaknesses of one method with the strengths of others.

Process improvements should be based on empirical evidence because it is only with the verification of assumptions, based on data, that real improvement will be realized. Table 7-1 illustrates a simple series of steps that could be used as the foundation of a CI investigation.

Table 7-1. Process improvement steps.

Improvement Step	Process Improvement Actions
1	Stabilize the process.
2	Characterize the stable process.
3	Determine the most significant issue for the process.
4	Determine the root cause of the most significant issue.
5	Make process improvement to remove (or control) the most significant issue.
6	Evaluate the effectiveness of the change.
7	Repeat step 1.

Stabilize the Process

As shown in Table 7-1, the first step of any improvement initiative must start, at a minimum, with a stable process. This is where the standardization aspect starts. This is not to say that the stable process must be a *good process*, just that it must be repeatable and consistent for all operators involved. As part of the stabilization, it will be necessary to understand the variation that is normal to the process from that which has a *special cause* and can be eliminated easily. The purpose of process stabilization is simply to remove the controllable variation in the process. Look at it as removing the noise within a process that hides the major problems that are affecting a process. This is an area in which MES can provide a significant benefit. When defined properly, the major configuration points (process routing, operator screens, and poka-yoke methods) within the MES for any product/process combinations will help to ensure that the process is both defined and followed.

To aid in process stabilization, the manufacturing engineer should consider some simple poka-yoke methods to ensure the product is fed consistently from operation to operation. For example, when connecting a wire harness with multiple connectors to the product, having the work instructions (and, of course, the operator training) call out placing the pending connectors in a specific location on the product carrier will help ensure that all connections are implemented properly. Following this procedure helps develop *muscle memory* (the expectation of a specific order of steps) and results in an extremely high detection rate when something is *out of place*, greatly minimizing missed steps. Each operation should be reviewed to determine the best product and material setup to invoke muscle memory, not just for the current operation but also for setting up the next operation in the process as well. Wherever possible, product jigs that are used to aid in the assembly process can be interfaced with the MES as a poka-yoke method to verify the activities.

Characterize the Stable Process

After the process has been stabilized, it will be necessary to characterize it. This step includes documenting the current process to gain an understanding of the normal variations and nonconformances, in addition to understanding the magnitude of their occurrences. Documenting the operating parameters of workstations and production equipment used within the process is also important. In many companies, there is an attempt to fix multiple issues within the process at the same time. The result is the effects of different changes possibly interfering with each other and giving false feedback on the cause of a specific problem—not to mention causing a large increase in labor to try to track the effects of each problem. This is the epitome of trying to "boil the ocean." The data recorded by the MES provides details about the process characteristics that are not possible to obtain without the system. In addition, as this data is recorded as a normal part of the production process, the MES provides a consistent representation of the process characteristics over multiple iterations of improvements. This provides validation at each iteration that the previous process changes remain stable.

Determine the Significant Issues

A Pareto report is a tool used to identify the *significant* issues and to provide guidance on addressing those that are of the largest magnitude (depending on what is measured). Using an interpretation of Pareto's rule that 20% of the problems cause 80% of the failure costs, Operations can focus on improving the issues that fall within that 20% to gain the greatest return on investment for the efforts of improvement.

Another method of focusing on issues that will achieve the greatest benefit is the TOC. According to the TOC, the focus should always be on improving the throughput of the constraining operation. In many processes, for example, a functional test operation is frequently determined to be the constraint. In this scenario, improvement may include increasing the capacity of the test operation or reducing the fallout of failed test cycles. Anything that improves the throughput of the constraint operation is a move in the right direction, and using a combination of the TOC and guidance from Pareto reports would have a significant impact on improvement. The key is to properly characterize the process by using data from a large enough sample that can be provided via an MES database and thoroughly analyzing that data to determine which issues are the most significant issues and which to resolve. Part of a company's Industry 4.0 initiative can be to use machine learning algorithms with MES data to create categorizations that can be used to build Pareto reports.

Determine the Root Cause of Issues

To determine the root cause of an issue, it is necessary to rely on empirical data. When viewed from an MES perspective, a process configuration that has been derived to form the stable process should be maintained as the primary configuration for normal production orders. As part of the activity to determine the cause of problems, the manufacturing engineer must engage in a DoE process to isolate the process characteristics that have a significant effect on the process issues being investigated. In this investigation, the manufacturing engineer must update the equipment configuration and the MES to collect the data needed to prove the causation of the problem. In many industries, the processes that have been changed for DoE can still be used to produce production units. However, additional validations within the process may be necessary to ensure usability. Because the MES has control and visibility of all active products, the system can be used to ensure that units being handled for a specific DoE initiative are quarantined from the remainder of the production units until validation is complete.

Make Changes to the Process

After all parameters of the issue have been evaluated, it will become clear which changes have the most significant effect on the problem being investigated (if not, it is necessary to keep looking or develop a control). Caution should be exercised not to change multiple parameters of the process at the same time. This will ensure that any changes made will not interact with other process parameters without detection. As previously stated in this chapter, a common fault of CI programs is implementing too many changes that affect downstream operations without having a holistic view of the process. From the MES perspective, once the parameters causing the problem are known, the original process should then be updated and implemented as the new standard process. Any product of the old process that is still in production (including in repair or other quarantine) should become the focus in order to cleanse the production environment of old products. This is not only good operations management from a material management perspective, but it also ensures that the data from old processes is not confused with production runs under the new process. Again, the MES can provide visibility of where the old products are located to clear production for the new process.

Evaluate the Effectiveness of Changes

Most CI programs fail to take the additional steps required to ensure the implemented changes achieve the expected improvement goals. The benefit of using an MES in your CI analysis is that the data required to compare prechange performance to postchange performance is automatically collected and available for comparison at any time. By data mining in the MES database, the manufacturing engineer can compare multiple runs of a product whenever required.

Once the new process has been implemented in the main production stream, it is necessary to stabilize the process again and recharacterize it to start the next CI initiative.

7.3.3 Product Improvement

Several concepts can be examined for product improvement, many of which are beyond the scope of this book. The primary concept we will address in this chapter is DFM.

In the scope of DFM, there are several areas of product design that can create problems for manufacturing. In this section, we will highlight some of these issues and discuss how an MES can help clear them up.

The following is a list of some of the issues faced within the guidelines of DFM.

- **Component location on the product** – This applies to the placement of components or subassemblies onto the product that makes it difficult to assemble or does not provide access clearance for rework or testing.
    - When an MES is used to control the actual assembly process and is configured to be aware of the processing time of each step, assembly times with significant variance for the same component can be interpreted as flags for investigation.
- **Material usage** – Using certain materials within the manufacturing process may add a level of performance for the product but may complicate the manufacturing process, increasing the cycle time of production.
    - In this context, a balance must be met between *necessary* performance and ease of manufacturing.
    - An MES provides the product engineer and the manufacturing engineer with empirical data that helps achieve this balance.
- **Qualifying material** – This requires a managed and repeatable process that enables the direct comparison of material under evaluation and excludes process variations that could affect the qualification process.
    - Using an MES as the control for process execution provides that managed and repeatable process.
    - After a test run of a material, the data collected by an MES can be reviewed by the product engineer to verify that there were no issues that would affect the comparison.
- **Design verification** – Like qualifying material, design verification requires having a managed and repeatable process to ensure that a new product or product version is performing as expected.
    - During a ramp-up for a product or a revision introduction, many companies identify problems with design configurations that were not seen during early validation processes. Experience indicates that many such issues found during "ramp-up" can be traced back to variations in the manufacturing process that gave skewed results during earlier verification runs, or as a result of confusion with some of the special handling requirements provided to production during these earlier design verifications. These issues will cause last-minute delays in product or version introductions.

- Using MES can resolve both of these issues by keeping clear visibility of which special handling instructions (and data collection procedures) are required by specific orders and by providing the same repeatable process management required in DFM.
- Chapter 8, "New Product Introduction: Details and Benefits," addresses new product introduction processes and provides more discussion on this topic.

A common concern facing many original equipment manufacturers (OEMs) is determining when to release a new version of their product line. The issue they face is justifying the engineering and support costs of releasing the version with the benefits gained by the new release. However, the production cost of reworking the old revision to deal with a problem that the new revision solves is frequently overlooked.

The costs of releasing a new version of a product vary greatly depending on the product and the industry. However, the cost of release is a one-time cost. This cost must be compared to the cumulative cost of performing any rework activities on the old version until the new release is ready. Getting a detailed understanding of the rework costs can be difficult. If the rework has been defined within the MES, the actual *touch time* of production operators can be much easier to calculate and can be made available at any time. With the MES, it is possible to know who touched the product, for how long, and how frequently, providing an accurate cost analysis.

7.4 More on Quality Methodologies

In CI management, the key to a successful program is to be methodical. Whether a company develops its own methodology or uses some (or all) of the industry-recognized methodologies is not as important as the methodical improvement of process and/or product over time. In addition to improving the product/process, the effectiveness of the manufacturing control systems and methodology should also be revisited. Every methodology will have pros and cons; this is inescapable. Knowing the pros and cons of each methodology used will help maintain a focused program of CI for years. Using an industry methodology for the sake of advertising is a sure way to make the delivery and effectiveness of the program a disaster. When methodologies are used in this manner, they are only implemented half-heartedly and will have a short life span within the company (until the next buzz method comes along). This manner of implementation will also waste a considerable amount of money and create just as much frustration for the staff as they deal with *yet another* method of implementation.

Implementing an industry-recognized methodology for CI quickly provides a roadmap of procedures and software tools that will kick-start the CI program. As part of the implementation of these industry methodologies, a company must understand how to use them and their weaknesses.

The next section presents an overview of CI methodologies and includes a review of their basic concepts and the strengths and weaknesses that have been expressed over the past few years. All these methodologies have been extended well beyond manufacturing and have been used to help companies in many industries gain considerable improvements in operational performance. A detailed explanation of these methodologies is outside the scope of this book; however, a considerable number of in-depth books are available for each methodology.

For this overview, we will examine the methodologies from the context of manufacturing only.

7.4.1 Six Sigma

The purpose of using the Six Sigma methodology is to identify process improvements that will invoke a major step in the effectiveness of the manufacturing process. The primary concept of Six Sigma is to use a rigorous set of management and analysis tools to drive all the characteristics of a process (and the performance of a product) as close to the target characteristics as possible. For example, where there may be an acceptable temperature range for a curing oven, with Six Sigma, it is not acceptable to maintain that temperature anywhere within the acceptable range. Six Sigma would require an understanding of the correct target temperature and would perform an analysis to determine why the temperature of the oven drifts away from the target temperature.

After the causes are identified, effort would be applied to identify which cause has the most impact on that drift, and then a change to the oven or process (depending on the cause) would be implemented to minimize that drift as much as economically possible. Finally, the process would be reanalyzed to ensure that the oven is performing as expected after the change is stabilized.

This is, of course, considerably simplified, but the premise of this methodology is that if all aspects of a process that cause drift in performance, or defects, are controlled to the most economical level and maintained as close to the *target* as possible, then the process has the lowest likelihood of creating a defect. This, in turn, helps ensure good performance of the product over its life span.

Six Sigma provides several methods of analysis and the necessary tools to perform those analyses. In addition, the Six Sigma methodology also defines the management processes needed to ensure the Six Sigma program is consistently managed over the long term. If a company is not interested in reinventing the wheel by creating its own methodology, Six Sigma is a well-established improvement management program. Six Sigma management is a good tool to use as a foundation for continuous improvement.

A significant issue with Six Sigma is that it is rigorous and detailed, and if a company is not serious about changing its culture, it can be difficult to keep a Six Sigma program running long-term. In the early life of the Six Sigma movement, it was not uncommon for people to become certified for Six Sigma (Black, Green, or Yellow belts), which would require them to implement an actual Six Sigma project during their training and then, after training was completed, they would not use the methods and tools for the remainder of their time with the company. This issue was a result of companies wanting to use the Six Sigma initiative as a marketing call and not really looking at it as a change in their management methodology. Without the proper support from management, the Six Sigma program would eventually fade.

7.4.2 Lean Manufacturing

The purpose of using the Lean methodology is to determine small and ongoing incremental process improvements in the effectiveness of the manufacturing process. Like Six Sigma, the Lean initiative concept is to provide the methodology and tools needed as a foundation for the company's CI program. The primary concepts of Lean can be summarized into two key goals:

1. **Voice of the customer** – Analyze the processes the company uses to deliver its product or services. Determine the steps within those processes that add value (from the customer's perspective) and then systematically remove any steps that would be considered a waste (again, from the customer's perspective). The main driver of this concept is to ask (and even specifically ask the customer), "Is this step something the customer would be willing to pay for?" and if it is not, then figure out a way to remove or minimize it. As a result, any aspect of the company's operation that does not provide something the customer is willing to pay for would be considered as waste (wasted labor or wasted material). Action would then be taken to eliminate or reduce this waste using predefined methods and tools from the Lean framework.
2. **Supply the customer on demand** – The product should be available to the customer only when the customer asks for it and only in the quantity the customer requests. If the product has not been ordered by the customer, there is no reason to be working on it. The result of this approach is that, theoretically, a product should not be released to work in process (WIP) unless it is directly linked to a customer order, and it should be made to satisfy the customer

order as quickly and reliably as possible. Additionally, the target process should have no need for buffering and zero waste (from scrap, rework/repair, or excess inventory). To move toward this goal, the company acts to remove the need for buffering WIP (having a balanced line) with all operations having approximately the same cycle time and staying within the estimated takt time needed to deliver the product when expected. The company also acts to remove the possibility of lost time because of equipment failure and to manage line changeover to new products, which includes equipment setup times that target single-minute changeovers. If equipment is not needed for a product run, then it is to sit idle or to be used by a different product during that time. At no time should a product be worked on anywhere in the system if it is not directly tied to a customer order.

Also, like Six Sigma, the validation of waste removal is required after each improvement project, and Lean provides tools to guide management in determining which improvement projects are needed and how to keep the Lean initiative going long-term.

When examined from a manufacturing perspective, an issue with Lean is that it is arguably not suitable for some industries. For example, when looking at a seasonal product like lawn chairs, a single company is not likely to have the facilities to provide the market with enough product to satisfy an on-demand manufacturing delivery for an entire season. Companies that make lawn chairs must estimate the demand well in advance and start production well before a single order is received. In this scenario, using some Lean concepts would not be as fruitful. It could be argued that sales could locate a different market that would provide a more consistent flow of product, but as lawn chairs are a highly local market, this is not likely to be of much benefit for the manufacturing floor. Whether or not a company could benefit from Lean in the rest of the enterprise is beyond the scope of this book. Industries such as the automotive sector, alternatively, have made excellent gains by using Lean concepts. This is to be expected as Lean originated from the Toyota Production System.

7.4.3 TOC

When looking at the TOC strictly from a manufacturing perspective, it is another operational methodology with a framework of procedures and tools used to help Operations clarify the goals of a process, define the limiting factors of that process (called the *constraint*), and then take two distinct and concurrent directions:

1. Maximize the throughput of the constraint through various actions.
2. Make process improvements that increase the throughput of the constraint.

The primary driver of the TOC is that within any process (or system), there is something that is a constraint or limiting factor. The constraint within a manufacturing process defines the throughput of the entire process.

The key understanding of the TOC is that regardless of the activities that occur elsewhere in the process, improving the constraining operation is the only way to improve the throughput of a line. Furthermore, if time is lost at the constraining operation from downtime or improper scheduling, that time is lost and is not recoverable. As a result, effort should be made to schedule material to the floor and any preceding operations to the constraint to ensure that the constraining operation is scheduled and loaded to its most efficient capability. The rate of flow through the constraint should be the driver of all other activity (referred to as the *drumbeat* to which production follows). Any preceding operations should be scheduled to ensure enough of a *buffer* in front of the constraint so it can continue if other operations are shut down (e.g., due to scheduling demands). Additionally, only enough material that can be processed through the constraint in a scheduled period should be pulled onto the floor. This process is like a rope pulling the material as needed, giving rise to the term *drum-buffer-rope* to describe TOC scheduling concepts.

The other key to understanding the TOC is that only activity that improves the throughput of the constraint will improve the bottom line. If an activity is undertaken to improve the capability of an operation or a piece of equipment that is not the constraint, it will be a waste of money and time because it will do nothing (or very little) to improve the throughput of the process or system.

As a methodology, the TOC also has analysis tools that are used to identify and highlight the constraining operation and to validate the effectiveness of changes to the constraining operation.

Similar to Lean, some production environments may be deemed *not suitable* for the TOC. The Lean concept of a balanced line could be interpreted as not having a constraint. However, it could be argued that a balanced line is simply making *time* the constraint, and all other rules and methods still apply.

7.4.4 DoE

In all previously described CI methodologies, there is one common factor: the concept of empirical evidence of an issue within a process or operation being the primary controlling factor. In other words, because of all the factors that may influence a problem, proof is needed to determine which factor is the primary influencer.

In the methodologies described previously, DoE is the primary tool used to isolate and prove that factor of influence. DoE provides a method of analysis that requires the manufacturing and quality engineers to determine all possible factors that influence a problem and create a series of test scenarios to isolate each factor. Next, each factor is varied to determine which has the greatest influence and, therefore, which is the highest priority to be controlled. The same methodology is then used to prove that by controlling the factor, the process is under control.

7.4.5 Failure Mode and Effects Analysis and Process Failure Mode and Effects Analysis

Failure modes and effects analysis (FMEA) and *process failure mode and effects analysis* (PFMEA) are product and process planning tools used to highlight ways that a product or process may fail (the failure mode); the likelihood of the failure occurring; and if the mode of failure occurs, its effects on other materials, components, or operations within the product/process. When this information is understood, the FMEA process then looks at ways to (1) detect this mode of failure and (2) prevent the failure either from occurring in the first place or from causing further damage or problems, thereby controlling the impact of the failure mode. As with other CI initiatives, the FMEA (or PFMEA) has a set of procedures and tools that is used to help define the issues and create the controls.

7.5 Additional Notes

This book does not advocate or rebuke the capabilities or methods of the methodologies described in this chapter. Instead, the authors encourage Operations management to understand them and use them as tools to control and improve their companies' efficiency in making products.

A primary component of all these CI initiatives is the use of data to prove the need to make changes to processes or products. In addition, each CI initiative advocates methods and tools to be used to stabilize and control the manufacturing process. Reducing variation in a process improves the capability of making products error-free. It also makes it easier to identify and isolate the issues causing variations (and, therefore, defects). When configured correctly, an MES can significantly help a company's CI program regardless of the initiatives and methodologies used. The key issue is to ensure the correct use and implementation of an MES to support the goals of a CI program.

At this point in the book, you likely have noticed the recurring theme regarding MES: in addition to leading operators through process steps and ensuring process control,

MES also collects data and correlates that data to the events that occur during production. The key to ensuring that an MES can provide the support discussed in this book is to understand how to configure the system's master data so that the data is readily available and the integrity and interpretation of that data are protected. In addition, a CI effort must include periodically revisiting the MES environment over time. As a company progresses in its CI program, the demands of data collection and process stability and control will change as well. Ensuring that your MES continues to accurately represent the production environment is just as important as improving the process and product.

7.6 CI Benefits

In summary, the benefits of using an MES in CI include the following:

- Provides the data required to analyze and validate the issues that can cause defects
- Establishes a platform for controlling and maintaining stability within a process
- Ensures a stable platform for testing variations in processes in a controlled manner

8 New Product Introduction: Details and Benefits

8.1 Overview of New Product Introduction

It is important to understand that the new product introduction (NPI) environment is a function that includes just about all aspects of the company. Product Engineering, Marketing, Finance, Sales, and Manufacturing all have a stake in a product that is going through the NPI process.

Questions that Operations should answer include:

- What will the unit cost of the product be?
- What is the demand-to-capacity ratio of the company's production or the production of one plant?
- How fast can the product be available to the market, or what is the *time to market*?

Even if the NPI process is just handling a new revision of a product, all these questions still apply. Costs may change, which provides a better price point within the market, or capacity may change, which provides greater fulfillment of demand and is possibly tied with marketing to increase market share. Many of these factors are also directly related to decreasing a product's time to market.

One of the greatest problems with NPI is having the product ready for initial shipment reliably. Significant preparation goes into launching a new product (or a new revision of a product), and planning the first shipments when marketing gets started is very important in any market. As is evident in the smartphone industry, some companies delayed the initial release of a product and lost considerable market share, while other companies released a product that had problems, knowing that the issues would be found, but at the consumer's expense.

As a result, it is critical that a company has the tools to move a product through the development phase as quickly and reliably as possible. Although most of the management of product and process design information is handled by product lifecycle management (PLM) during the NPI process, there are specific concerns regarding managing the process when products and processes are evaluated on the production floor.

During the design phase, it is important for materials to be tested to ensure performance characteristics are met and to make design and supplier decisions. In some industries (e.g., the automotive industry), many materials must meet regulatory requirements before being considered for use within a product.

Preparing for material tests may include partial or complete builds of the product at its current level of design in order to perform statistical testing. In these scenarios, the product build must be highly controlled because any deviation in the build process between materials or designs being analyzed can affect the testing of the product or material. This can cause the product engineer to analyze skewed data or a product that has built-in, hidden inconsistencies that may affect the evaluation. In many cases, the skewed results may not be detected until late in the product development phase (if at all before product release) and cause unexpected delays in the release date, or worse yet, a company may release the product anyway and suffer the consequences to their reputation.

Anyone familiar with design for manufacturing (DFM) principles understands that Manufacturing Engineering must be involved in the design to (1) ensure the product can be manufactured efficiently and (2) prepare the manufacturing line to build the product at an acceptable rate of production to supply the expected demand. The preparation of the line may include the design and build of production-level test capabilities and the creation of assembly jigs that will be used to ensure repeatable manufacturing.

If the product assembly is highly automated (as is the case with many components in the automotive industry), it is not just the jigs that must be built but the automated fabrication tools as well. Having Manufacturing ready to produce the product in expected quantities is critical to the successful launch of the product (or revision).

Getting the timing right when introducing a product can make the difference between a successful launch and capturing a major segment of the market, and an unsuccessful launch that misses a marketing opportunity completely, potentially causing a company to go out of business. The high-tech industry has some excellent examples of the ramifications of having troubles during NPI.

In the next few sections, we investigate the areas of NPI, how an MES can help bring products to market faster, and how the correct configuration of the MES is crucial to an effective NPI business process.

8.2 The Operations Processes of NPI

Within the scope of operations, a manufacturing facility must develop an understanding of several issues as early as possible in an NPI project. These issues include the cost of the product, the supply of material, and the capacity of the available manufacturing facilities, all of which must be forecasted well in advance. In addition, each of these areas of concern contains multiple subissues.

8.2.1 Material Cost

Although the cost of the components sometimes makes up a significant percentage of the product cost, most of the component cost tracking is managed by the supply chain or procurement as part of a vendor management program. Using an MES provides visibility to the components that are being used in the build process. In many cases, the bill of material (BOM) does not contain all the data on what was consumed because it only applies to the material for the initial build. Material will also be consumed as repair and rework are performed. As products are sometimes processed and reprocessed through operations, an MES tracks all steps executed and maintains a record for product and process engineers to use when evaluating new designs. An MES can provide a clear understanding of the additional material that was consumed during these processes.

It is important for any vendor management program to include input from an MES to get a clear profile of a supplier's performance. Product engineers can use MES reports to highlight components and subassemblies that require more work to increase reliability. The next section discusses this further.

8.2.2 Transformation Cost

The goal of manufacturing (or transformation) is to convert raw material into a product by following a fabrication/assembly process. The more efficient the process is at transformation, the lower the cost. This enables a company to lower prices in a highly competitive market or increase profit in less competitive markets. For efficiency, the process is designed to have only the steps required for assembly (i.e., operations that add value), and instructions for assembly and the controls are made visible to ensure the operations are performed correctly. To decrease the time to market, the process development function must complete the design in as little time as possible, and the design must be lean in order to maximize the yield and minimize the amount of repair or rework required.

If you look at the product development process, you will notice prominent functions in the product development lifecycle where an MES can be of significant value. Once the product engineer understands the information that an MES provides, they can use the system to quickly verify that a component is *fit for use* and recognize product features that make it difficult to manufacture. Adjustments in product design can then significantly reduce the manufacturing lead time of a product. For more information, investigate the subject *design for manufacturability*.

8.2.3 Material Evaluation

As part of the product evaluation, using components that are known to be reliable can greatly reduce the touch and wait times (waiting for repair/rework) to manufacture a product. Ensuring a supplier is delivering good components requires a good supplier relationship.

Best-in-class supplier relationships will minimize the number of suppliers used by a company to gain the best price and provide the most consistent forecast for the company's suppliers to plan against. The problem is determining which suppliers to use. Some of the information used to identify a good supplier comes from the audits of received goods. If a good incoming material audit program is in place, a company has all the information required to make that decision, right? Well, not quite.

Material and components from suppliers may be received in good condition but may not be capable of surviving the production process. These failures may be due to a bad process or a bad supplier. The NPI process should be capable of testing these components in a *production-like* environment. Developing the manufacturing process on the same MES that runs production provides the ability to profile a process by recording detailed data from repairs and process parameters. This method helps with understanding the stresses placed on the material during production, including whether the stresses are a normal part of the manufacturing process (and therefore must be tolerable to the product) or are due to a special issue that will require a change in the process. Having this data made available in an MES gives the product and manufacturing engineer actual data to review and greatly clarifies the source of problems.

Using an MES to define and control the material evaluation process within NPI (or as part of a general incoming inspection methodology) enables detailed product (or subproduct) tracking and provides the control to ensure the product (or subproduct) has been made consistently. This method provides good statistical data for product engineers to evaluate performance.

As for these materials, tracking failure information from testing and repair is the only way to get the full picture of a good supplier. Even if a component makes it through the production process, a component from a bad supplier may fail shortly after delivery to the customer. Getting the repair information from the MES for both production and return material authorization (RMA) is a vital part of any supplier rating function. Using an MES for both your production repairs and RMAs provides a holistic view of information on supplier performance and can be a significant input to attaining a stable product and process in a shorter amount of time, decreasing a product's time to market.

8.3 Planning Capacity

The demand for the product, to put it simply, is a forecast of sales for a market. As such, calculating the expected demand is beyond the scope of this book. However, understanding the capacity and the drivers for capacity as they are reflected within the NPI environment is a major part of operations planning.

While a product is in the NPI process, significant work must be done in the planning phases of a product launch (into full production) to determine the company's expected capacity. Failure in planning a product launch can lead to unsatisfied customers who are not able to get their orders fulfilled. Just as bad, a company can overestimate its required capacity and invest in manufacturing facilities (or contract manufacturing facilities) only to have these facilities underutilized, causing the company to fall short of its return-on-investment goals.

Whether a company is planning for a greenfield facility or upgrading an existing brownfield facility, a major part of that capacity planning is determining the process for manufacturing and estimating the expected yield of processes. In this section, we will investigate some of the various drivers of information needed for capacity planning and how MES can help with that planning.

8.3.1 The Manufacturing Process

The function of NPI in the context of the manufacturing process is to prepare and prove the product for full-scale production or develop the manufacturing process profile for planning purposes. This may involve making many changes to the process or test parameters, updating the work instructions, and running the process with specific component batches for test purposes. As previously noted, companies can also use the NPI process as part of their supplier qualification processes as well. As a result, the definition of a "stable process" for manufacturing (as in most manufacturing environments) is not likely possible.

Moving a product through NPI may also require running many different versions of the product configuration or design in small batches through multiple versions of testing protocols. Doing so ensures that when the product moves into *full production,* it can be manufactured efficiently and cost-effectively at relatively high volumes.

With all these changes in product and/or process, it is difficult to track the current versions that must be maintained. There are also times when certain combinations of process and product should not be part of a preproduction run. Tracking all these changes manually can be a nightmare for any Manufacturing Engineering team.

An MES works so well with NPI requirements because of its ability to track and control the changes in product and/or process. An MES application performs all these tracking functions as a normal part of its design. The manufacturing engineer can incorporate controls in the process to ensure the status of certain product versions are available for NPI production runs or are set to be obsolete and reject any activity for these products. With the visibility of material batch IDs included as part of the data collection process, an MES is also well suited to verify correct material use before the material is consumed during the preproduction runs. By using an MES for process management in this manner, manufacturing and product engineers can be assured of having the data for the correct product or process version during their analysis.

When combined with a good MES strategy, including policies on operation, resource ID format, and routing rules, the MES application provides a strong platform to quickly define and modify processes to work with any NPI requirement.

8.3.2 Platform for Process Definition

When it comes to the manufacturing process, it is essential to have an application platform that provides easy tools for guiding (and controlling) the flow of product and displays work instructions (WIs) to the production operator at the right time in the process. A good MES will not only display the WIs but also track if (and when) the operator has viewed them.

By using standardized templates of processes and WIs, a product or manufacturing engineer can quickly define changes to a process, determine the WIs needed, and request specifically trained production operators to perform the work, all within the normal MES functionality. Standardized process development should also include testing, nonconformance handling, and reporting.

When the manufacturing engineer has defined an NPI line that has more generic operation and resource definitions (which will prepare the line to handle the wide diversity of NPI process requirements), configuring the MES application to recognize and support that line will take a large burden off the production operators for line changeovers and promote a production environment that greatly reduces the variation that causes human error.

When the interfaces between the MES and the automated manufacturing equipment or test equipment have been standardized, the communication, data format, and performance reporting make developing and commissioning a production line, and even new production equipment, much faster and, more importantly, much more predictable.

An additional extension to the standardized interfaces is consistent data that is used for reporting. One of the more difficult issues with material analysis and process comparisons is obtaining reports that can be interpreted equally. When the data is collected and reported consistently, the analysis and interpretation of reports for product design and *process capability* are much easier to perform.

8.3.3 Using Alerts and Event Detection

When product and manufacturing engineers run evaluation samples, there are times when the actual results of the sample are badly out of spec. Often when this happens, the test or inspection results can quickly indicate a problem in the production run. But without real-time visibility of the results, the production line running the evaluation sample has no indication that there is a problem until the sample run is complete. By using an MES, deviations from common characteristics can be detected early, and an alert can be automatically sent to the responsible engineers. This gives the engineers an opportunity to make no-go decisions early in the sample run, saving time and money for the NPI project.

8.4 Additional Notes

When implementing an MES to support an NPI environment, those developing the strategy must recognize that the key to successfully using an MES is to configure the MES master data into highly modularized components and to develop and maintain standardized templates for defining processes, work centers, reports, and WIs.

Manufacturing Engineering is responsible for defining and testing many process segments within NPI. Therefore, in NPI, there is no "stable process" to maintain. The program strategy, master data, and configuration policies must be defined in a manner that provides building blocks for each part of the MES environment. When defined properly, the MES application can be an excellent tool for product and manufacturing engineers to use to determine the correct product or process design and for material selection. When an MES is utilized at the start of NPI (to be implemented and operational at the start of production validation), better efficiencies in product design and production definition can occur, realizing cost benefits to the manufacturer and greatly improving the speed and reliability of NPI.

8.5 NPI Benefits

The benefits of using an MES with NPI covered in this chapter include the following:

- The MES application platform can define the templates used for process design and WI creation.
- An MES provides a controlled environment to collect and report data/metrics used for product and/or process analysis. This is useful during production ramp-up as the manufacturing processes are validated.
- The MES application platform can manage manufacturing the large number of concurrent revisions for both products and processes used to test acceptable performance capabilities.
- An MES provides a controlled environment to generate data to use for capacity planning in forecasting NPI timelines.
- AN MES can validate system readiness at the time of production and support ramp-up and process validation instead of conducting system validation at post-production ramp-up.

9 MES Architectural Concepts

9.1 Introduction

Over the past several chapters, we have reviewed various business processes and the data required to develop efficiencies within the manufacturing enterprise. This is the reason *why* a manufacturing execution system (MES) is valuable. However, we have not yet reviewed the *how* of MES—how does Operations use an MES to execute the functionality of each process and provide these efficiencies? This chapter will cover these items.

The following topics will be covered in this chapter:

- General architecture (programmable logic controllers—PLCs, servers, databases, supervisory control and data acquisition—SCADA, applications, etc.)
    - Plant-floor devices (scanners, sensors, etc.)
    - PLCs and data concentrators
    - Plant-floor human-machine interfaces (HMIs)
    - Servers, clients, and distributed computing
    - Cloud, edge, and fog computing (all external to the plant)
- Industry 4.0 and the Industrial Internet of Things (IIoT)
- In-plant (internal hosting) computing with edge versus cloud versus fog solutions
- Autonomous MES

While it will not be possible to cover all the possible permutations of architectural solutions, the basics will be covered so that system architectural details are understandable.

9.2 General Architecture

Most MES architectures follow the ISA-95 hierarchy model (covered in Chapter 2) on the different levels of manufacturing activity (see Figure 9-1). In breaking down the levels into system architectural elements, Figure 9-2 becomes clearer.

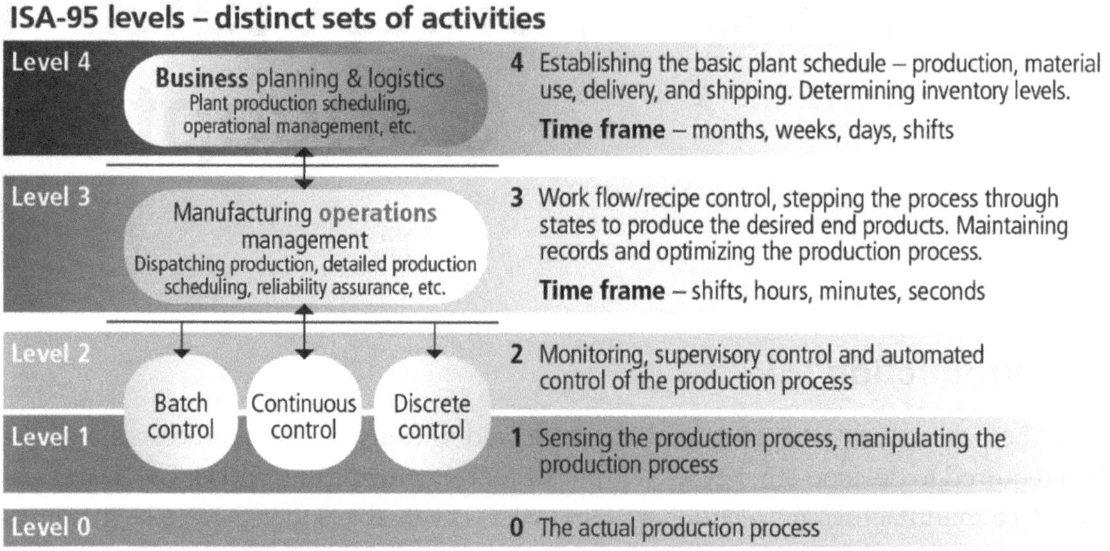

Figure 9-1. ISA-95 hierarchy model levels.

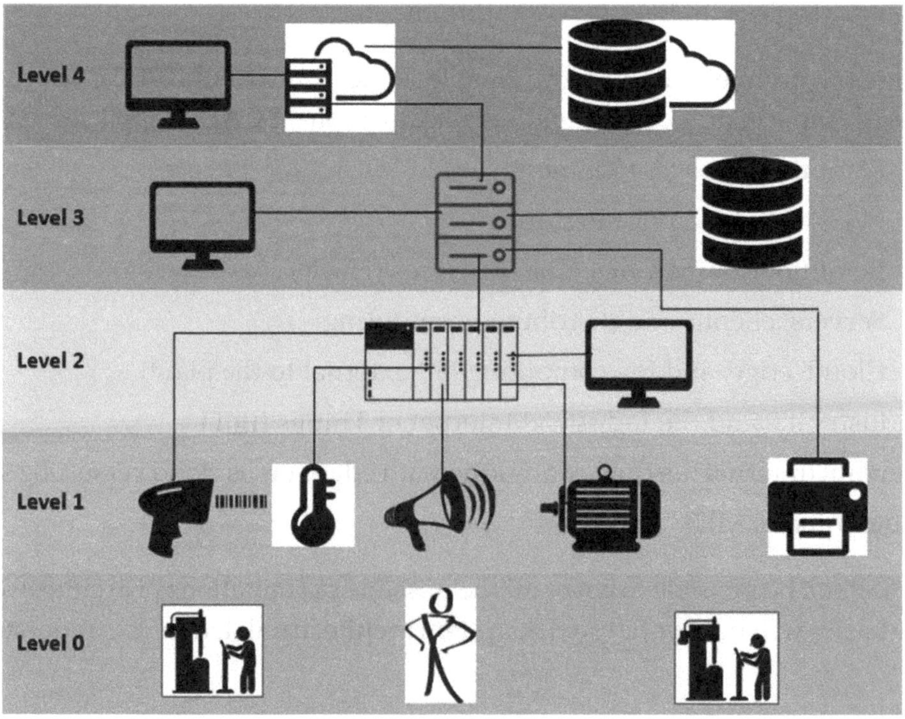

Figure 9-2. System dataflow in an ISA-95 hierarchy model.

The following becomes apparent:

- **Level 0** – This level represents the actual work being performed. This includes the tools and automation that act on the product itself. Examples include inserting components and bending tubes, shafts, or presses.
- **Level 1** – This is the first layer of monitoring that senses the actual product manipulation; scanners and annunciators are Level 1 devices. They are used to generate the data for SCADA machines.
- **Level 2** – This is the controls layer. It is made up of PLCs, data concentrators, and other data computational devices, and system monitors used to provide supervisory controls or to display automated controls of the manufacturing process. These include any dedicated closed-loop or small process controllers (not shown in Figure 9-2). This level requires Levels 0 and 1 to control the process.
- **Level 3** – This is the operations layer. All supporting processes required to manage controlling the manufacturing process, in general, are located here. Servers, relational databases, and workstation clients are located here. They help manage the process through screens, reports, and other business functions directly associated with manufacturing. Real-time interactions that occur between multiple operations within a process are frequently managed at this level. Most of the business logic that is used to maintain the integrity of both the processes and the product being manufactured is managed at this level. Manufacturing data is provided to this level from Level 2 or directly from Level 3 operator screens.
- **Level 4** – This is the business planning and logistics layer. All long-term analytics and production planning are done here. All data is provided from Level 3. This includes all business functions that support but are not directly part of manufacturing, such as customer relationship management (CRM), supply chain management (SCM), and other functions. This is the level that cloud-based solutions can help support.

It is important to clarify the types of clients that are present at each level. At Levels 1 and 2, the clients are the HMI devices that provide the ability to interact directly with the production equipment within the controlled processes. These devices are functioning and connected directly to the equipment controllers: interaction here must occur in real time, so there cannot be an information technology (IT) solution that causes a delay in the communication interfaces. The interfaces include setting up registers in the PLC for the annunciation of production faults that require an immediate response and scanners that read data from the product to the PLC or SCADA systems.

At Level 3, the system's clients interact with IT systems to provide data entry and system configuration support in addition to system status monitoring, reporting, and other dashboard-type functions. These clients will interface with the system servers and, in certain architectural solutions, with the database that provides the relational storage of system and production data. Most of the IT interface functions for an MES will bridge Level 2 data to specific functions of Level 3.

Like Level 3, Level 4 includes clients that provide access to IT functions. The difference is that Level 4 does so at the enterprise level above the MES. Depending on the specific system, this includes functionality from enterprise resource planning (ERP), SCM, CRM, and product lifecycle management (PLM). It also includes production reports and comparisons over multiple plants or long-term trending and analytics. This level is not concerned with (and not recommended to be) a direct controls interface (Level 2). As a result, the functions here are mostly no longer in real time but are based on database functions that may require cascading transactions (as seen in Chapter 2, Figure 2-3). As previously mentioned, this is the operational level that could benefit the most from cloud computing solutions. While we are not going to cover these computing solutions in depth, we will cover the solutions in order to understand their impact on an MES and system responsiveness.

- **Cloud computing** – This refers to the application functions processed on server facilities that are typically external to the company's domain and provide computing resources (central processing unit and database resources) on demand. Some cloud providers can also support on-demand resource allocation on computing environments that are internal to the company's domain, known as *internal clouds.*
- **Edge and fog computing** – This computing solution is a hybrid of cloud and in-plant computing. Its main purpose is to limit the amount of communication going out into the cloud, thus minimizing issues with usage bandwidth on the Internet. Edge (or fog) provides data compression and data preconditioning (calculations and data manipulations) before requesting transactions from the cloud. However, although fog and edge are frequently used interchangeably, in reality, fog extends the cloud environment into limited general processing to domain internal servers, while edge only provides the data compression and preconditioning.

The point to consider is that each solution has its advantages and disadvantages, the impact of which will be covered later in this chapter.

9.3 Industry 4.0 and the IIoT

Over the past few years, there has been considerable buzz about the terms *Industry 4.0* and *IIoT*. Both of these terms allude to the same thing and represent an IT influence on MES architecture: the ability to access and use vast amounts of data that was not previously available.

Let us quickly explore these terms. *Industry 4.0* represents the inclusion of computers and IT solutions into the industry/manufacturing environment. It follows the other major developments in manufacturing:

- Industry 1.0 refers to the use of different types of power in manufacturing. This dates back to the early 1800s and included the eventual use of electricity to assist in manufacturing. Prior to this, all manufacturing was manual labor and tools. General skilled labor (i.e., skilled in the entire manufacturing process) moved to where the product was being assembled rather than the product moving to the labor (labor skilled at specific operations at specific operational steps).
- Industry 2.0 refers to the use of conveyance systems in manufacturing. Before Industry 2.0, conveyors, parts, and tools had to be brought to the product. With Industry 2.0, conveyors, parts, and tools could be assigned to dedicated workstations, and the product was brought to each workstation. This change decreased the overall production time, which in turn decreased costs and increased product quality. Industry 2.0 dates back to the late 1800s, with the most successful and influential implementation being the Ford Motor Company in automotive manufacturing in 1913.
- Industry 3.0 refers to the use of computing systems (PLCs) in manufacturing to automate the individual manufacturing steps. Each function of automation was independent of other activities on the line. Before PLCs, manufacturing controls were a set of hardwired relays executing control only at the bit level. Troubleshooting operational issues and making changes were cumbersome. PLCs made controlling the manufacturing process easier. Industry 3.0 dates back to the late 1960s.
- Industry 4.0 employs computers, networking, and other IT solutions to fully digitize and connect all manufacturing equipment. All data from one machine

(called a *cyber-physical device*) is made available to and can directly influence other machines. This capability was always available through MES in a limited fashion, but Industry 4.0 takes it to a higher level and MES is already a major part of it. Although there was some discussion of the concept earlier, the formal concept of Industry 4.0 was originally proposed at the Germany Trade & Invest event in 2011.[1]

During the age of Industry 4.0, controls and data features of plant-floor devices have continued to increase to the point where end devices are now fully networked. End devices (e.g., sensors and actuators) not only have some computational function to support controls, but they can also provide a direct interface to networks for parameter and live process data. This is the activity around the IIoT. While the Internet of Things is concerned with this growth of ability across a wide spectrum of solutions, including vehicle and home monitoring and control, the IIoT is concerned directly with the activity in and on the manufacturing plant floor.

The impact that Industry 4.0 and IIoT concepts have had on manufacturing cannot be underestimated. Not only are more devices and hardware required for full system functionality, but also more data interfaces are needed to support communication. This, in turn, represents more points of failure in system functionality. The additional level of data may or may not be valuable in an organized MES operation. Careful determination of the requirements for a manufacturing operation (from data to hardware expense) must first be determined. A small manufacturing operation will not have the resources required to implement and support a big system.

9.4 In-Plant (Internal Hosting) versus Cloud versus Edge or Fog Solutions

Possibly the largest concern regarding IT is the cost of ownership—not only the cost of purchasing and setting up a computing infrastructure (server and client hardware, networking hardware, etc.) but also the software and operating systems (operation system configurations, firewalls, other security configurations, etc.). All items have nontrivial expenses associated with them. In addition, once computing resources are configured, they are dedicated to a function regardless of the demand for that resource. Given this, cloud solutions were developed to provide IT solutions, enabling

1. German Trade and Invest is an annual event where German industries discuss changes in trade and technology investments that can benefit German industry as a whole.

the plant to leverage a pool of computing and storage resources that are hosted off-site as a better cost model. These off-site facilities provide variable computing resources as demand varies.

Let us dive into the differences:

- **Internal hosting** – The plant is responsible for all IT operations. The advantage is that the entire data model is not public and is fully managed by the manufacturing organization—any changes to functionality or support can be responded to quickly because it is in-house. The disadvantage is that—depending on the size of the organization—this can be an expensive endeavor.

- **Cloud** – The advantage is that the plant requires only limited personnel to maintain IT hardware, security, or upkeep. In addition, as the system resources are made available "on demand," the plant only has to pay for the services used. The disadvantage is that the cloud depends on the health of the wide area network or Internet (which has outages and delays based on weather and external bandwidth usage); as a result, it may have large fluctuations in system response time. The plant must rely on the provider to secure client-sensitive data as well. Depending on the capabilities of the provider, this may, in fact, be a benefit because some providers will have extensive security capabilities (much better than most in-house networks) to protect their cloud environment. An additional concern about the cloud is that the programs (i.e., the MES applications) are shared between several customers. As a result, frequently there are limitations to the number of changes that can be made to the MES application, and therefore, a possible limitation to the support that the cloud MES will provide to the production floor.

- **Edge or fog** – As mentioned, this computing solution is a hybrid of cloud and internal hosting. The advantage is that moving non–real-time sensitive business functions off the in-plant hardware and onto the cloud minimizes the IT impact on the plant, thus providing cost savings and maintaining the processing speed needed to support real-time transactions. Disadvantages may be recognized with the increased complexity of a company's networking requirements and in support of the manufacturing IT systems.

9.5 Autonomous MES

Autonomous MES is based on what can be considered the next step in computing, called *agent-based computing*. The concept of agent-based computing has been available in test cases for various topics since the early 2000s. A white paper on using this technology for MES and how autonomous MES could work was introduced to a discussion forum in 2018 by Grant Vokey of Vokey Consulting Inc. Since the introduction of the white paper, there have been a few improvements in technology that can help make this concept achievable. The following is an excerpt from the white paper as an introduction to autonomous MES. The original white paper is included in the Appendix, "Introduction to Autonomous MES of Computing."

9.5.1 White Paper

Introduction

In recent years, the computing power of devices has grown considerably, and that computing capability has been made available in smaller and smaller packaging to the extent that the common functionality of a smartphone now has greater computing capability than a stand-alone computer from a couple of decades ago. In addition, more diverse communication paths have now been made available within the packaging of a smartphone device. These communication paths include Wi-Fi, Bluetooth, and cellular and near field communication (NFC), all standardized within a single package. With some redesign, additional communication paths can be developed, including designs in spread-spectrum FM communication.

By using devices similar to smartphones and the capabilities being developed in IIoT devices, the introduction of a new MES infrastructure is now possible. By using these computing devices (referred to within this document as *nodes*) and the development of small programs to perform very specific functions (referred to as *services*), it is now possible to create a new MES infrastructure using mostly nodes to perform the specific tasks needed within the MES environment by designing these nodes to support peer-to-peer communication. Because these nodes will only be supporting a very narrow scope of functionality, the speed with which these tasks can be performed will make it easier to provide the real-time speed of communication needed within an MES.

As shown in Figure 9-3, the proposed implementation of this new infrastructure would use peer-to-peer communication between specialized nodes for the majority of transactional communication. Within this infrastructure, the nodes engage with each other directly without communication to a central system. The result would be a significant distribution of the necessary communication to subsections of a company's network, thereby reducing the demand for bandwidth within the network that is currently seen in conventional MES infrastructures.

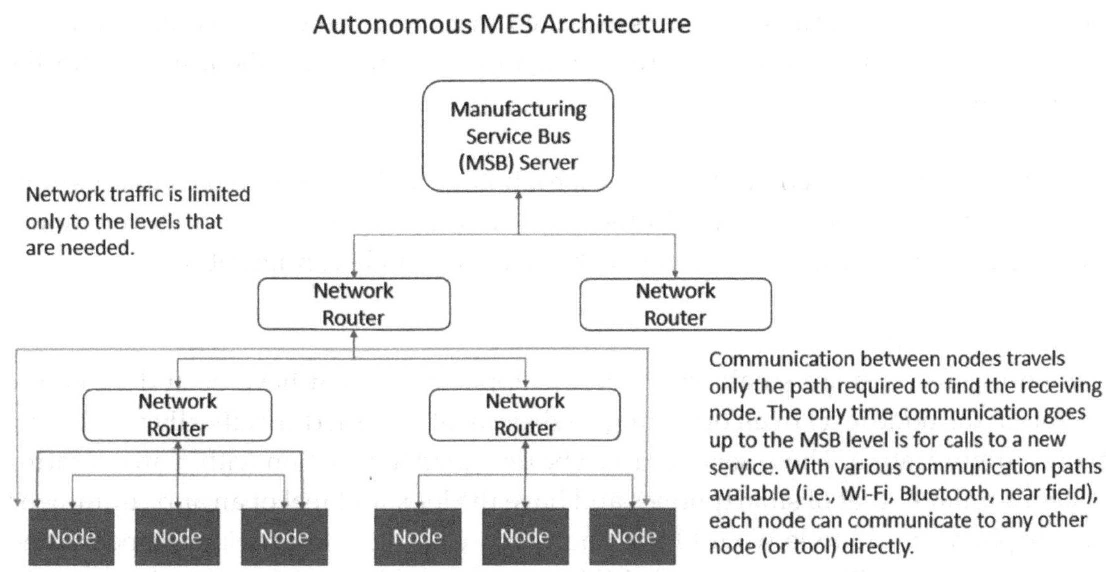

Figure 9-3. Network communication—autonomous MES.

Components of the New MES Infrastructure

Manufacturing Service Bus

The manufacturing service bus (MSB) is the only part of this concept that maintains the centralized computing structure. The purpose of the MSB is to provide a repository for services to the nodes and to maintain and process the MES activities that are not of a transactional nature. These MES activities would include functionalities such as reporting, data aggregation, and event management. Further description of these functions will follow later in this document.

Computing Nodes

Computing nodes are small computing devices with the capability to communicate over multiple communication paths, such as Wi-Fi, cellular networks, Bluetooth, and NFC infrastructure. These nodes could also be designed to handle Ethernet communication, if required. Further options for communication may include other spread-spectrum communication protocols yet to be developed.

The purpose of a node is to act as an autonomous computing device that is mounted as an integral part of equipment on the production floor, such as a workstation (the location on the production floor where work is actually performed), product carrier (equipment or containers that are used to move units that are in production—WIP [work in process]—through the production process), or line-side stocking (LSS) unit (locations used to maintain component inventory needed for the production of units). Within the concept of the autonomous MES, it is expected that there could be possibly hundreds of nodes within a production environment, all performing specific tasks within the MES as a whole.

As part of configuring computing nodes, each node will be required to maintain its own IP address, its purpose within the MES architecture (i.e., what MES function it will be performing), and where it fits within a master node configuration.

Computing Services

Computing services are small, very specific applications that have been designed to operate at the node level in an operating environment designed specifically for peer-to-peer communication. These services must be designed to function within an operating system similar to that of smartphones and have the look and feel of an app running on a smartphone. Each node would have one or more services to provide the peer interaction necessary within any specific MES function. Additional functionality must be embedded into each node to provide the capability for each node to communicate with other nodes over the various communication channels.

These services would also include application support to communicate with other equipment within the defined communication protocols. This may include communication with test and sensory equipment, automated actuators, or other tools needed to perform specific tasks.

Each service would maintain the data locally (on the computing node) that is specific to that service for the duration of the time that a production unit is considered WIP. Once that unit is complete (either to finished goods or scrapped), all of the data for that unit would be aggregated and transferred to the central data warehouse relieving the services from the responsibility of holding that data.

Some of the tasks needed in normal MES functionality would include the reporting, data aggregation, and event management, introduced earlier. The services to provide the necessary support for these functions would operate within the MSB infrastructure, making them available for use within a cloud environment.

MES Assembly Interaction

In this interaction, the services that will be participating in this example (referred to as *actors*) will be the carrier node and the workstation node. Each would have been preconfigured to operate with the required services when the nodes were originally placed into service. The workstation node would also be configured to communicate via Bluetooth or an NFC sensor (part of the smart sensors talked about earlier in this document). As an option, the workstation node would also be configured to communicate with an LSS node located at the workstation, and the workstation would have a display node also configured for the operator.

When the NFC sensor detects the carrier node, the sensor triggers the workstation node to engage the carrier node; through the NFC sensor, the carrier node and the workstation node perform a handshake and initiate communication via Wi-Fi.

Once Wi-Fi communication is verified, the carrier node recognizes the operation that the workstation node is to perform, and the carrier node provides the workstation node with the list of materials required to assemble (including part count and possibly data collection requirements).

At this point, several interactions could take place:

1. The workstation node communicates with the LSS node to ensure that the required material is available.
2. Then the workstation node communicates with the display node to inform the operator of the list of materials required to assemble.
3. The LSS node shows the operator which material to assemble first (using pick-to-light) and indicates the assembly sequence for the remaining material.
4. As the operator picks each piece of material and assembles it, the workstation node communicates to the carrier node and to the display nodes to update the list of assembled items.

Once the assembly process for that workstation is complete, the workstation node and the carrier node perform a final handshake to ensure all data has been updated, and the two nodes end their peer-to-peer communication. The carrier node will then be moved to the next workstation to perform the next step in the process.

If the assembly process included the use of tools, each of the tools would be preconfigured to the workstation and could communicate with that workstation over Bluetooth or peer-to-peer via Wi-Fi for intelligent tools.

9.5.2 Additional Notes for Autonomous MES

Technologies that have been developed since the original publication of the white paper could also be used within the autonomous MES structure. For example, block chain technology could be used at the carrier node level to maintain the data for an entire production order to ensure data consistency and security should any of the nodes fail. A new node could be made available and updated with the correct information from within the block chain structure.

Another option is to use fog capability to extend the MSB into the level of nodes close to the activity that is being monitored (i.e., close in cyber proximity).

9.6 Additional Notes

9.6.1 Utilizing the ISA-95 Standard

The technology that supports ISA-95 by an MES varies considerably between software vendors, and the implications of the lack of support for a specific technology also depend greatly on the environment in which it is implemented. Knowing the options that are available for each implementation and how to translate them into solid solutions will be a major part of the decision-making on an MES strategy. To help develop a solid MES strategy, many companies are using the ISA-95 activity model as a template for process requirements that an MES must support. Using the activity model as a template helps to ensure that the full support of manufacturing operations management is covered. Note, however, that not all MES providers support or even recognize the importance of the ISA-95 standard as it pertains to an MES strategy. If a client company has little growth expectation (either organic or merger and acquisition) as a result of industry characteristics, the support of ISA-95 will make little contribution to the long-term benefits of any particular implementation.

For most (if not all) companies, growth is of primary importance. As a result, during a company's existence, there is a high likelihood that either the ERP system and/or the MES must be replaced as a result of changes in business demand or changes in technology. By becoming familiar with the standard and looking for vendors who support it, a company can optimize its flexibility to change the system as its needs change.

9.6.2 Interfacing with Enterprise Applications

For many companies, interfacing between the enterprise level and the MES can be an issue. Figure 9-4 shows a traditional situation that can be a considerable problem with existing interfaces between enterprise-level applications (e.g., the PLM and ERP) and the MES. The PLM–ERP interface usually predates the implementation of the MES, and companies are highly likely to resist changes to this interface. In the example of the PLM to ERP, when an MES is implemented in the architecture, there is a strong tendency to continue using the PLM–ERP interface to import data from the PLM through

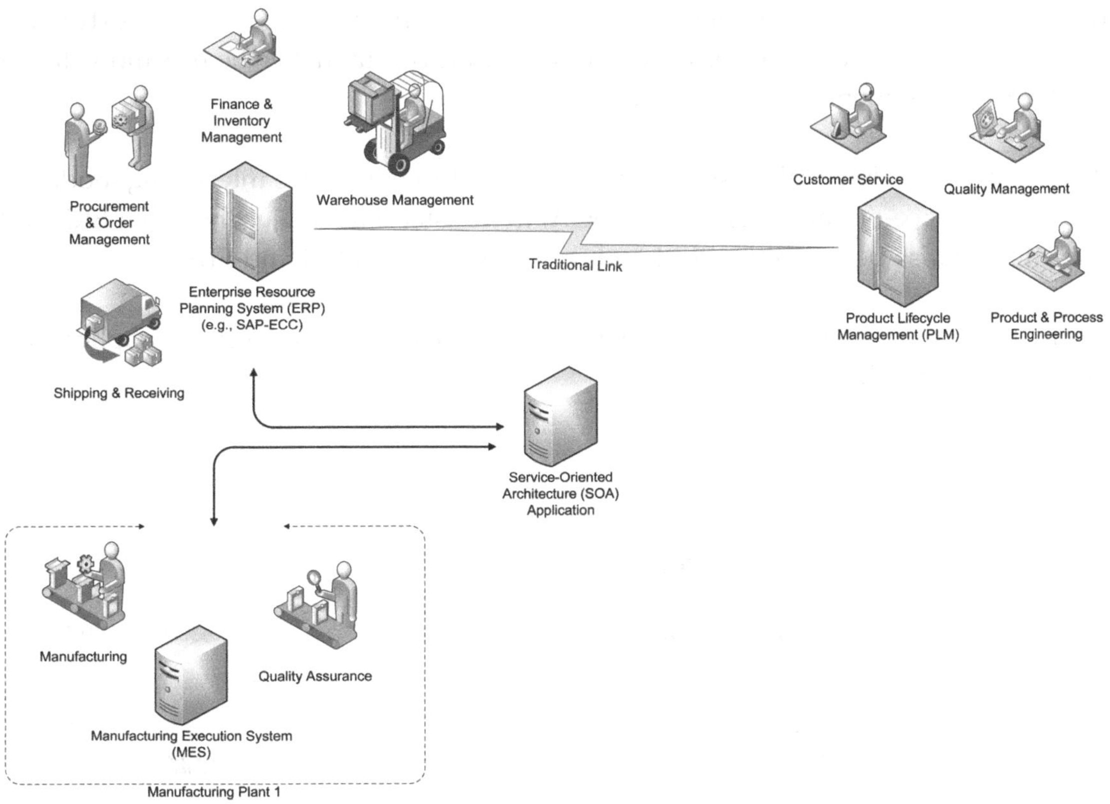

Figure 9-4. Traditional enterprise-to-MES interface.

ERP to the MES. This requires considerable changes to the PLM–ERP interface as well as internal changes to ERP in order to handle the MES-level data that is not typical to ERP (and for which it has no purpose).

This type of interface also makes the upkeep of the MES difficult as any changes to the MES requirements force additional work to be performed on the following:

- **The PLM–ERP interface** – To update any data transfers that are required by improvements in MES capability.
- **The ERP system** – To handle additional data objects that must be temporarily stored until being transferred to the MES. This data may need to be permanent to handle data integrity investigations.
- **The ERP–MES interface** – To handle the transfer of data that is intended for the MES.
- **The MES** – To maintain the data that is intended for the MES.

In this traditional environment, MES updates and maintenance become extremely costly and, in most cases, cost prohibitive. This, in turn, stagnates the use and value of the MES environment.

A few companies have taken steps to improve the full enterprise interfacing model by using the service-oriented application (SOA) to transfer data not just between ERP and the MES but also between the PLM and the MES and the PLM and ERP.

Figure 9-5 shows the interfacing model between the enterprise layer (all applications at that layer) and the MES.

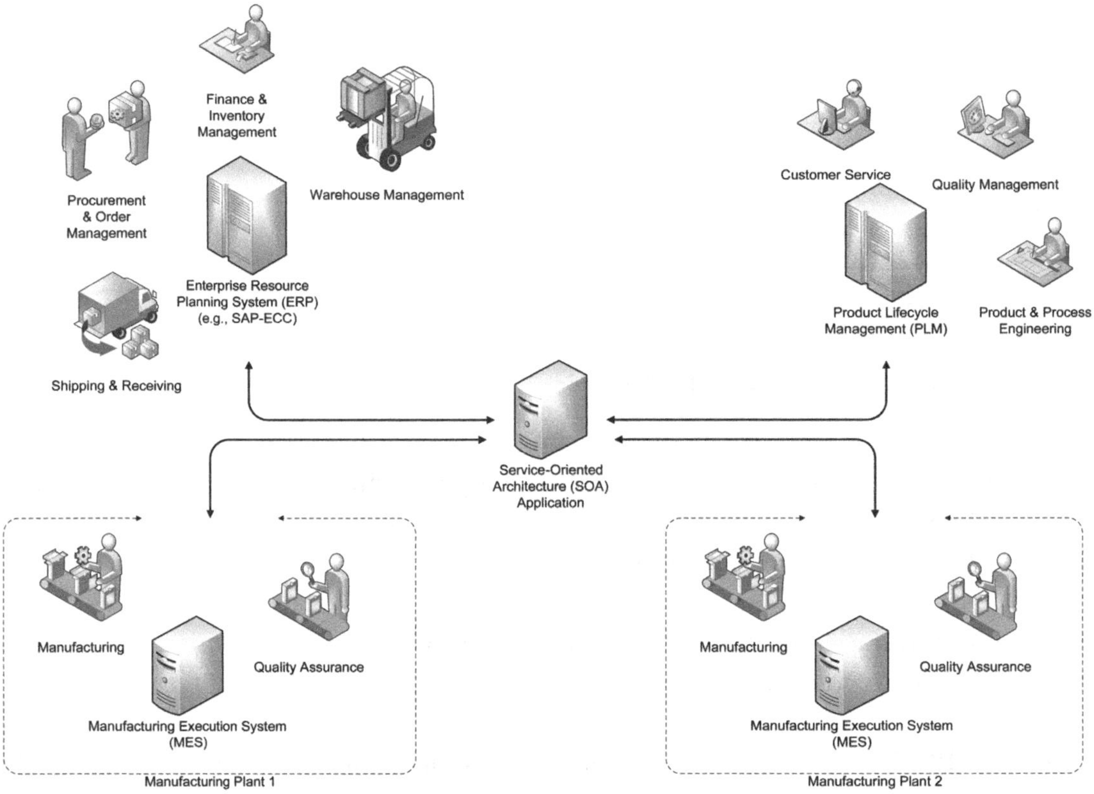

Figure 9-5. Enterprise interface to the MES with SOA.

By using the SOA layer for ERP–PLM communications as well as PLM–MES communications, the interface is capable of making data transfers much more efficient by:

- Exporting data from any application and splitting it as required to other applications that are connected (e.g., export a PLM bill of material and import the planning extract to the ERP and the assembly extract to the MES)

- Making updates and maintenance much simpler because only the required sections of interfaces must be updated
- Making the data models simpler because each application has data modeled for only what is required by that application
- Adding additional MES (for other plants), which only requires modeling of the one interface to that MES
- Simplifying adding, removing, or changing any of the individual applications after the original interface is created

This level of interfacing will make the initial implementation more complex to handle the original design, but in the long term, this interfacing model is much easier to maintain.

9.6.3 Interfacing to the Shop Floor (Equipment, PLC, and SCADA)

As described in Chapter 2, the primary concern of interfacing with the production floor is to ensure that data is made available and processed quickly.

The problems that affect many MES implementations are:

- The diversity of data structures made available from ISA-95 Level 2 to communicate
- Determining which level is going to be the controlling layer of the process (Level 2 or Level 3)
- Determining how the data from Level 2 is going to be represented at Level 3

At the PLC/SCADA level, the primary data recording function is a data historian. These data storage systems are designed to record detailed parametric data that is available as a data stream. As a result, historians are designed to record one specific measurement from the data stream and mark each record with a time stamp as the key index of the data record. The issue with analyzing data from a historian is that if you are searching for process characteristics in relation to other events on the production floor, you have to know the time stamp of each parse of the data stream that would be of interest. In other words, if you are determining the temperature of an oven for any production units, you must first determine the time that each unit went through the oven, parse the data stream from the historian for that time element, and then plot the normalization curve (as another example) for each production unit. If you are doing this analysis for more than a few units (consider a sample size of 100 units of production), you must repeat this for each production unit before you compare them. If the SCADA system has the entry time of the production unit stored in a different record of the historian, SCADA systems have analysis tools that can make this a little easier. However, if multiple events are being analyzed at the same time, this process can become extremely complicated.

When using an MES, it is possible to sample parametric data when the production unit enters the oven (in the previous example) and record the values directly to the unit's device history report record. As the MES application has a record of most events (users, equipment, components/subassemblies, operations, etc.), it becomes far easier to analyze the relationship between multiple events.

However, the relational data of the MES is not a good storage location for the detailed data stream that a historian would provide. As a result, it is important to "condition" the data that is parsed from the historian to compress the detailed stream down to a few data points (e.g., average value, ±3 sigma deviations) before storing the data within the relational database. The correct conditioning technique required for the data depends significantly on the characteristic being recorded.

In some cases, the SCADA system may be designed to parse a section of the historian data when a trigger has been received from the MES and then precondition the data before replying to the MES trigger. Whether preconditioning before replying or sending raw data is required depends on the characteristic being recorded. Planning the recording of data, whether in the MES or the historian, requires coordination between the controls engineering staff and the MES support staff.

9.6.4 Configuring the Database

An additional part of the MES configuration is the structure of the database itself and the reporting that will be used from the database.

In most cases, the database used to support the transactions of the MES from day to day will be structured to provide quick access to data from the perspective of the unit identifier (serial number or batch ID) and/or the production order in the form of a *star schema*, as shown in Figure 9-6.

As a result, all other data (material master, bill of material, routing, workstations, etc.) must be related to one of the primary data objects (product ID, unit ID, and/or production order). (Data objects are sometimes called *business objects*, or BOs.) A *star schema*

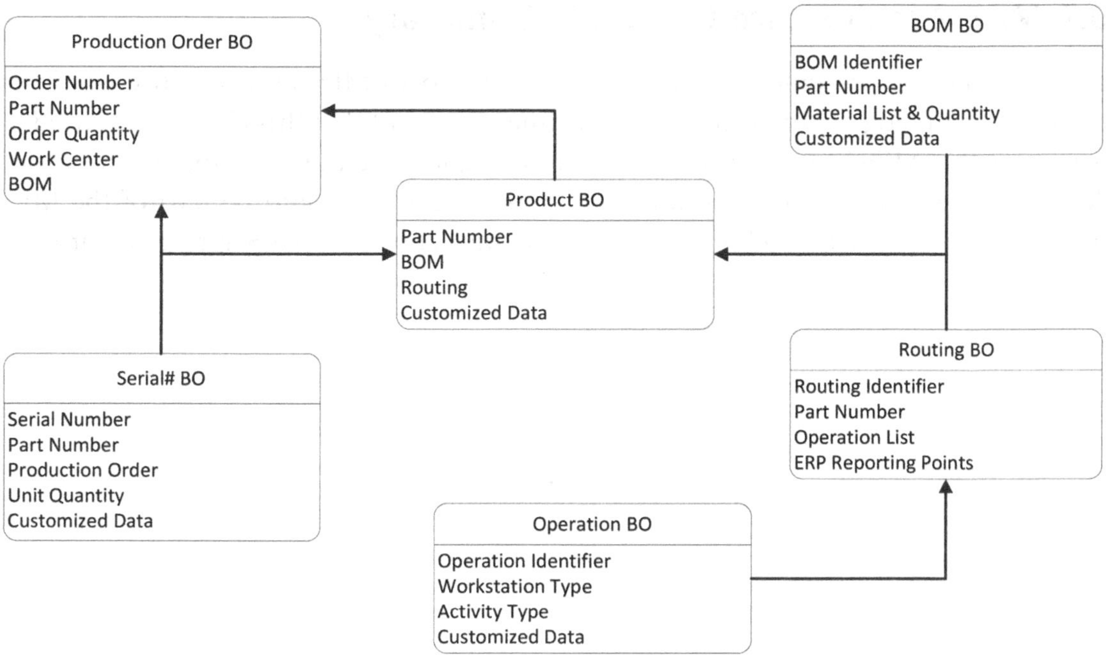

Figure 9-6. Star schema example.

structure is designed to have complex relationships and provide quick access to a specific context of records. If it were not designed in this fashion, the application would find it hard to maintain the "real-time" transactional tracking. These schema structures are also good at generating the reports used by production staff, supervisors, and dashboards and, as a result, may also be referenced as the *operational database.*

Reports that are frequently required from this schema are status-type reports that tell Operations staff what is important about the work being performed at that point in time. These reports are usually very specific in regard to the information that they convey.

The star schema structured database, however, is not effective when reporting information that is related to trends in either the product or process performance. When these reports are needed, the data from the operational database must be changed to make trend analysis easier. For this type of reporting, a data extraction (from the operational database) process is used, in which the data records are "flattened" and stored in a data warehouse for longer-term storage. Data that has been extracted from the operational database can then be archived after a period of time as this data will likely not be needed by Operations after the production order has been completed and shipped.

9.7 MES Architecture Concepts Summary

This chapter has presented several concepts related to MES and has discussed the trends of technology being utilized. The primary concept that this chapter has tried to express is that MES is much more than just the implementation of a single application. As a result, a considerable amount of strategic planning should be part of the MES implementation, much of it before discussions start about the MES application itself.

10 Differences in Manufacturing Models and Their Impact on MES

10.1 Introduction

Now that we have covered the *why* of manufacturing execution systems (MES) through the review of business processes and the *how* of information technology (IT) solutions, we must review the different models of manufacturing to best understand how each type has its own MES characteristics. Without doing so, it would be difficult to configure an MES properly. In this chapter, we will cover these differences and how they shape MES operations. First, we will cover the different types of manufacturing and explain their characteristics:

- Discrete
- Batch
- Continuous
- Hybrid

We will then apply the manufacturing types to examples of manufacturing models and explore the main manufacturing industries and their primary requirements for an MES:

- Automotive
- Aerospace
- Food and beverage
- Metallurgy (aluminum, steel, etc.) and plastics/polymers
- Pharmaceutical
- Others, including tires and general consumer goods

We will then align the main MES functional areas to the industries to complete the picture.

10.2 Manufacturing Models

In the following sections, we will review some characteristics of various manufacturing models and provide an overview of important MES functionality that is required to support these models. This will provide guidelines on what functionality to look for when investigating MES solutions.

10.2.1 Discrete

In manufacturing, *discrete* means that each production unit produced is independent of the others process-wise, and there is no real connection between one unit being produced, its predecessor, or the unit that comes after it. Therefore, each production unit can have its own specific path through the manufacturing process. As a result, the production tracking and management must recognize each production unit as its own distinct entity within the MES and must use a serialized unit identifier.

Because each production unit may not exactly match another unit, all elements of the MES (i.e., product scheduling, subpart sequencing and line delivery, quality assurance, and production monitoring) must recognize and handle each production unit independently.

When process management can reach high levels of first-pass yield (the ultimate target is a Six Sigma process capability), it then becomes possible to test and inspect production units at audit levels instead of 100% of the units in process. When this level of process capability is achieved, production units can be arranged into groups called *process lots* according to their similar process characteristics (e.g., same material, equipment, and operators). The process lot is then *sample tested* to determine the *fit-for-use* status. This is similar to batch management with the exception that the process lot is only managed for the length of the testing cycle for a particular process lot. Companies that are targeting this level of production management should look for an MES capable of handling process lots.

10.2.2 Batch

The batch manufacturing model is used when many production units follow almost exactly the same process and use exactly the same material and resources. As a result, there is very little variation in quality and handling from unit to unit. In this model, the processes for manufacturing a product are frequently very simple from a management perspective, or each production unit is so inexpensive that it is not worth managing at a *per-unit level* (there are, of course, exceptions such as pharmaceuticals, where the unit price within a batch of drugs can still cost thousands of dollars). In these circumstances, the manufacturing schedule is set up to produce a relatively large number of units that are all almost exactly the same, and all units produced at that time are all identified by the same batch or lot ID. Because the entire batch of production units has similar quality characteristics, each batch is usually sample tested to determine its conformity to requirements, and the quality and processing records are recorded for the entire batch instead of at the individual unit level.

In process manufacturing, a batch (or lot) would be determined by the number of production units that are all made from the same batch of raw materials and by the same set of equipment and production operators.

From an MES perspective, the application must be configured to handle multiple units of a product under the same production unit identifier (batch ID) and have the capability to process partial quantities of units distributed across the entire process. In this model, the inputs into the production run (process, equipment, and material) are usually controlled to define a change in the batch ID (if any of the inputs change, then the batch ID must change). If a subquantity of the batch is redirected because of nonconformance or special processing, the application must be able to separate that number of units and reidentify them as a batch variation by changing the batch ID representing that number of units.

In this model, the data collection, quality control/assurance, and reporting of characteristics of the batch are usually performed by statistical representation and not at the unit level. There are going to be exceptions that must be handled within specific industries. For example, in pharmaceuticals, specific *production lots* may need serialization and special controls (double-blind unit tracking) to handle clinical trials.

10.2.3 Continuous

In continuous manufacturing, the process itself must run continuously in order for the product to be produced, even though distinct end products may be produced. While this sounds like a batch process, a batch process can be shut down at almost any time without affecting the ability to make the product. The best example of a continuous process is where the raw materials used in manufacturing are processed in some manner (e.g., heated and maintained at a certain temperature during most of the process) to be formed into a usable part. So, materials like aluminum, steel, plastics, and other polymers must all be maintained at a controlled temperature for manufacturing to occur. In oil and gas, the raw material is processed to produce a continuous flow of different co-products (gasoline, oils, kerosene, etc.). Any downtime in production must be scheduled, and any unplanned downtime must be avoided; loss of control of the processing environment (like cooling of material) will cause equipment (like pipes) to be damaged or, worse yet, personal or environmental harm.

In this type of manufacturing, process monitoring and quality assurance are the most important aspects of an MES. In this context, quality assurance refers to the reporting of statistical process control and process statistics on parameter drift as well as run-time statistics in control of process parameters.

10.2.4 Hybrid

Many products that are manufactured have subassemblies that have been manufactured in discrete, batch, or continuous methods. These subassemblies will be combined in some manner to create the end product, which requires a hybrid manufacturing method that uses more than one manufacturing model. In this manufacturing environment, each subarea of manufacturing must have an MES configured in such a way as to properly manage the specifics of that manufacturing model. Once the subassemblies are manufactured into the end product, the MES must then be able to manage the scheduling, quality, part and subassembly material management to the line side, and the ongoing production monitoring requirements in addition to the overview monitoring by enterprise resource planning (ERP) as part of the total supply chain management. This covers both completing the subassembly manufacturing process and the process of assembling the subassemblies into the top-level product. This manufacturing environment becomes multitiered only in manufacturing facilities that run all these manufacturing models under one roof. If the manufacturing occurs over multiple facilities, the MES can be configured for each facility with some data sharing as needed to the MES in the final end-product manufacturing site.

10.3 Manufacturing Industries

Now that we have explored the different manufacturing models, we must explore how they apply to different manufacturing industries.

10.3.1 Automotive

Automotive manufacturing varies across the globe and depends on how large the corporation is and how many plants there are. We are only able to cover some generalizations here. The automotive industry is characterized as high-volume, medium-option content manufacturing and uses a detailed just-in-time scheduling method. The typical manufacturing rate for original equipment manufacturers and Tier 1 suppliers is approximately 60 vehicles per hour, making for a quick manufacturing environment that is sensitive to negative production events. All automotive manufacturing is organized into specific processes, similar to those shown in Figure 10-1.

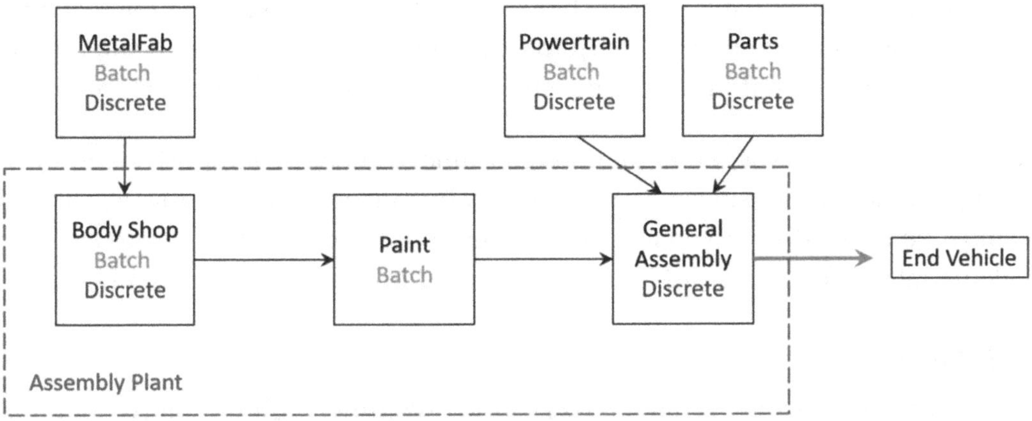

Figure 10-1. The automotive manufacturing process and flow.

For example, there are two starting points when a new vehicle is launched. One is metalfab, short for the operation metal fabrication, where the sheet metal body parts are first stamped out using batch processing and then partially assembled, which uses discrete processing. The other is powertrain, where all the parts and assemblies required to make the vehicle move are built, from engines and motors to gearboxes, transmissions/transaxles, and connecting shafts. Subparts like gears are built-in batch processes and then assembled in discrete processes. The sheet metal parts are then sent to the body shop area in the assembly plant. Here the vehicle subassemblies, like doors and fenders, are built into a batch process but can be tracked via serial identification.

However, as the vehicle body is built, the process then becomes discrete depending on the body options (two versus four doors, sunroofs, etc.). When the sheet metal of the body is fully assembled, the product is then transferred to the paint shop, and all bodies are painted in batch because it takes time and there are scrap costs involved with changing the color of a paint booth. After the bodies are fully painted, the vehicle is released according to a specific sequence into the general assembly where all remaining parts, including the powertrain, body frame, glass, interiors, plastic exterior parts, and all other parts, are assembled on the body. This is strictly a discrete process due to each vehicle having different option content.

Regarding parts providers, product manufacturing models depend on the parts being manufactured. Certain subparts can be built into the batch, such as speedometers, but once they are included in assemblies like instrument clusters and instrument panels, the process becomes discrete again because of the different option content. Also, typically these part suppliers are separate organizations, and the supply chain becomes multitiered, with Tier 0 being the vehicle manufacturer, Tier 1 being complete part suppliers like seat assemblies, and Tiers 2 and 3 being individual part suppliers like the aforementioned speedometers. Although one MES cannot be developed to manage the entire manufacturing supply chain, an MES can be developed for each manufacturing plant to ensure that any required hand-off data, such as genealogy, can be provided to the next plant in the supply chain in the form of advanced ship notice(s).

As regional and country legal regulations continue to evolve to support changing laws regarding public safety, especially in areas of collision and personal injury, all regulations must be understood to ensure the proper quality traceability and to ensure the genealogy data is not only stored but also reported in the prescribed manner.

10.3.2 Aerospace

Aerospace manufacturing is characterized as low-volume, high-option content manufacturing. Although only one airplane may be completed per day, each airplane can have a multitude of options, depending on whether the end customer is the military, an airline, a corporation, or a rich private citizen. Each end product can vary to the extent that one model of a production unit (an aircraft) may not be recognizable as being based on the same base model as another product. Because of this, aerospace manufacturing has a high content of batch processes for parts and discrete processes for assemblies throughout its whole manufacturing lifecycle.

Although *just-in-time* (JIT) is used in aerospace manufacturing, it is on a different time scale than in the automotive world. Whereas the automotive industry looks for JIT deliveries in the time frame of less than a day to make and deliver parts in sequence, aerospace has a larger delivery window but a much more complex transportation requirement in the supply chain. In aerospace manufacturing, greater importance is placed on safety and *fit-for-use traceability*, as well as a higher usage of statistical verification in batch management. This should not surprise anyone who recognizes that any failure in the operation of an airplane can cause a catastrophic event. As such, quality assurances, for not only the manufacture of individual parts but also the assembly process, must be tightly monitored and managed, with multiple levels of inspection and certification during the assembly process. A key aspect of aerospace manufacturing is validation that parts are *fit for use* and that only parts that have proven fit-for-use records are used during production. A similar requirement is also placed on the verification of production operators—either operators must be certified to perform the work, or the work must be verified by a certified inspector (the latter is referred to as a *buy-off*). One other element is to automate the work instructions (WIs) for the option content, which will also enable and automate the error-proofing functionality. As quality assurance has such significant importance, quality testing records are required to ensure all functions have passed the required performance specifications and safety thresholds to meet regulatory requirements. All these data points are then usually part of the traceability and genealogy records of the aircraft itself.

Regarding regulations, because the aerospace industry has global-based clients, most regulations are relatively common throughout the world or are recognized as supersets of either American or European regulations, especially in the areas of safety and personal injury. All regulations must be understood to ensure the proper quality traceability and to ensure the genealogy data is not only stored but also reported in the prescribed manner.

The focus of an MES in aerospace is quality assurance for traceability and genealogy due to regulations, and each assembly or test operation can be complex, leading to detailed information required at the user screen level. The order fulfillment and material handling functions of an MES are still important in this area, but screens have the added requirement of accessing WIs, computer-aided design files, and other functions specific to an operation. Although the manufacturing production schedule in assembly is not as critical as in automotive (although one could argue that bad scheduling of too many wing assemblies for an airliner would quickly become a critical problem), production monitoring for parts and process schedule is still important.

10.3.3 Food and Beverage

It should surprise no one that a certain amount of manufacturing is required to make, produce, and package food and beverages, whether it is simple produce or complex prepared meals. For safety considerations for human consumption, this whole manufacturing process (starting from the access of raw materials) must be closely monitored for cleanliness and sanitary conditions. This includes storage vessels, cooking equipment, packaging, and all handling apparatuses. Even the cooking process must be monitored to ensure proper temperature and times are used to prevent food contamination. Although there are continuous elements in food and beverage manufacturing (e.g., continuous ovens or fryers), food and beverages are usually processed in batches, and the processing is carefully monitored. Tracking the required equipment maintenance, from cleaning to upkeep, must be carefully managed.

Typically, production monitoring includes the proper scheduling and tracking of maintenance, cleaning, and upkeep. Closely tracking the genealogy of materials is an important part of an MES in this industry. Proper reporting of key areas, including genealogy and maintenance activities, must also be carefully planned.

Regulations can vary from region to region and country to country, but they are all specified to eliminate and manage contamination. All regulations must be understood to ensure the proper genealogy data and maintenance records are not only stored but also reported in the prescribed manner.

To support a more common management of regulations, most countries have now accepted the use of manufacturing management processes under the good manufacturing practice (GMP) or the current GMP (cGMP) process and equipment management models. Although many companies in food and beverage still use paper-based tracking, some MES applications have been *preapproved* to support the traceability and process control required by cGMP. Any company in this industry should be looking for this level of capability.

10.3.4 Foundry and Plastics

While at first glance the manufacture of metal products (e.g., castings, forgings, and the processing of ingots to sheet metal) and plastics may not sound similar, they share similar processes from a management perspective. All the raw materials for either metals or plastics must first be processed to eliminate impurities. Then they must be properly heated and mixed using the correct recipe to achieve the proper characteristics of the end product. The heated object must then be processed in a way to properly cool and shape it into the final product.

In these industries, continuous manufacturing is used almost exclusively to create the initial material, and then batch manufacturing is used to create an initial product. There may be some variation because some metals and thermoplastics can be reheated back into a liquid state, but thermoset-type plastics are destroyed when reheated. As with the continuous process discussion earlier in this chapter, monitoring production and managing the continuous-type process are key. Depending on the end product and the customer, quality assurance may also be required from a regulatory perspective, but there are industry standards; for example, some steel must be tested for proper density and strength characteristics as strength changes depending on how the steel is quenched. Other examples include testing the heat resistance of certain plastic parts and strength testing plastic parts.

For the most part, there are minimal regulations for metals and plastics, so industry-managed standards are used instead. However, there are cases with specific end products for which there are regulations for metal performance and plastics; as such, knowing how the metal or plastic product will be used is key to determining which regulations, laws, or standards are required for MES configuration.

10.3.5 Pharmaceutical

Like the food and beverage industry, the pharmaceutical industry's main manufacturing concerns are the purity of the materials, the sanitation and cleanliness of the manufacturing equipment, and the quality assurance of the end product, all of which are controlled using recipe management. All steps—the combination of the materials (chemicals) as well as the processes and resources (equipment and people)—used must be tracked and collected for genealogy. In this industry, the most important MES function is quality assurance, specifically the proper management of production activities that are used to regulate genealogy and reporting. This covers both prescription-specific pharmaceuticals and nonprescription (over-the-counter) pharmaceuticals.

Each country has well-defined laws and regulations for both MES data and reporting that must be followed, and violations can result in significant fines. These laws and regulations include what products can be sold and how each required report should be formatted. The MES must be able to support these configuration differences and invoke control over the process to ensure compliance.

As with food and beverage, the cGMP management model has become the primary management method industrywide in pharmaceuticals. Any MES used in pharmaceuticals must have functionality that supports cGMP reporting and management control.

10.3.6 Others

There are too many industries to cover completely. In this section, we will focus on tires and consumer goods. Tires are made using a continuous process to cure and process the natural rubber or synthetic rubber in a continuously heated process. The material is then bonded with either fabric or steel belts, and finally, the tread is attached. The rubberized material is a thermoset-type material, so it is critical to make the tire correctly before the material cools. The belts are typically made in a batch process, so the overall manufacturing process is hybrid. Because tires are the first product between a vehicle and the road, safety is a primary focus. Regulations can vary, but often the genealogy and traceability of the materials and process parameters are required as in other continuous processes described earlier.

Most consumer goods, such as appliances, electronics, oil and gas, pulp and paper, textiles (clothing), and wood-based products, are usually created using a hybrid manufacturing process and/or share most of the concepts that have been explained in this chapter. The regulations depend on the type of end product and can vary greatly. With these types of products, an MES can help to achieve process efficiencies—as described in earlier chapters on quality assurance and continuous improvement—to lower the operational costs of manufacturing.

10.4 MES Specifics Table

A summary of the manufacturing types and industries and how they relate to MES is provided in Table 10-1.

10.5 Summary

Understanding how products are manufactured is just as important as understanding the business processes of manufacturing in order to configure the MES and provide the best functionality possible. This makes for the last part of understanding manufacturing, other than reviewing the current set of manufacturing challenges, which we will do in upcoming chapters.

Table 10-1. MES function versus industry.

Manufacturing Item	Automotive	Aerospace	Consumer Goods	Electrical/ Electronic	Food and Beverage	Metals and Plastics	Pharmaceutical
Type							
Discrete	X	X		X			
Batch			X	X	X		X
Continuous						X	
Hybrid	X	X		X			
Order Execution							
Just-in-Time Scheduling	X	X					
Order Launch Broadcast	X						
Order Tracking	X	X		X			
Shipping	X		X	X	X	X	X
Material Management							
e-kanban	X			X			
Kitting	X	X		X			
Line-Side Replenishment	X	X	X	X	X	X	X
Subassemblies	X	X	X	X			
Quality and Traceability							
Error Proof/Poka-Yoke	X	X	X	X			
Test Plans			X	X	X	X	X
Trace/Genealogy	X	X	X	X	X	X	X
Work Instructions	X	X		X			
Production Monitoring							
Equipment Alarms	X	X	X	X	X	X	X
Production Alarms	X	X	X	X	X	X	X
Overall Equipment Effectiveness Metrics	X		X	X	X	X	
Throughput	X		X	X	X		X

11 Developing an MES Strategy

11.1 Introduction

To determine the direction for a manufacturing execution system (MES), several topics must be examined to ensure the right decision has been made at the corporate level. For companies that have multiple plants, getting this decision right can make the difference between satisfying the needs at the plant level with a total cost of ownership (TCO) that is effective at the corporate level or having pockets of effective MES use with an astronomical TCO without the benefits of a corporate-wide knowledge base. In many cases, the MES investigation starts with a single plant that is *ahead of the game,* but as each plant moves to MES implementation, it goes through the investigation phase on its own and with only the single plant's requirements in mind.

The cost of the first implementation is always the greatest as several initial interfaces must be developed and customizations are inevitable. In the situations previously mentioned, the implementations are scoped to support only one plant, and knowledge of using the MES is restricted to pockets at the plant level. In addition, each plant will frequently implement its own MES application (or develop its own), resulting in multiple MES applications being used, and configurations that should be common over several implementations must be designed and reconfigured in each separate plant. This also applies to customizations that should be common. Many large companies in the auto industry, for example, have already fallen into this predicament with the point of view that each plant is exclusively responsible for its own profit and loss and, therefore, its own investments in an MES.

Without an overall MES strategy, lessons learned are not easily transferable because each implementation is unique and may even be on a unique application, making it necessary for each plant to *relearn* how to solve problems that arise. Each time a plant must determine a solution to a quality or process issue, the solution analysis must start from scratch.

An outcome of this method of MES implementation is that plants that have independent MES strategies will not deliver products or information to the customer consistently, causing frustration when these customers are supplied by multiple plants. In one situation, a customer was receiving the same product from three different plants (all plants of the same supplier), and the customer's major complaint was that there was no coordination within the supplier's manufacturing organization. There were significant differences in product delivery that resulted in considerable frustration on the customer's part because issues that were solved in one plant needed to be revisited and investigated by the customer in another plant.

When a corporate strategy is put into place, benefits can be achieved by collectively determining requirements and implementing common and repeatable solutions. In this chapter, we will review many subjects that should be included as part of a full MES strategy. Even in situations where an MES is being implemented in only one plant, many of these suggestions still apply.

11.2 A Refresher on the Benefits of an MES

For manufacturing companies, there are three primary benefits to using an MES. The first is to provide a platform that gives companies the ability to stabilize their processes and provide work instructions that are needed by a production operator at the right time during the production process. This stabilization helps reduce variation and increase repeatability to prevent human error. When a solution has been determined, an MES provides a platform that permits changes to a process to be normalized faster. This helps speed up the implementation of continuous improvement initiatives. If a company has multiple plants that make similar products, an MES enables the company to develop and implement best practices across the entire enterprise.

The second benefit is that all activities in the production environment are recorded within a relational database. This makes it possible for quality and process engineers (and, in some cases, product engineers) to review what has happened on the production floor before having to go there to see it. When analyzing a problem, there is nothing better than actually seeing a problem in person (in Lean terms, "going to gemba"), but by using the database information, it is possible to narrow down where the potential issues causing problems are and reduce the number of gemba walks required to fully analyze these problems. Also, with the relational database of an MES, there is the added benefit of having the statistical data readily available to determine how much of a problem the issue really is (a major consideration for Six Sigma projects). These benefits improve the rate at which issues are pinpointed and isolated and provide for the development of specific solutions that resolve the root cause.

The third benefit (by no means the last) is that the relational database provides visibility of issues across the entire production environment. Many continuous improvement (CI) initiatives focus on improving a single operation and/or a single issue. One of the faults of CI initiatives that are managed this way is the inability to recognize that improvements in one area (or operation) may, in fact, cause problems in another. As an MES records the entire history of all production units, the impact of one change at an operation is easily seen in the output of another operation. This helps speed up the implementation of CI initiatives. If a company produces a product in multiple plants, and the plants are on the same MES, this can provide excellent insight and support for centralized improvement programs. In turn, this enables companies to develop and implement best practices across the entire enterprise.

It is possible to accomplish these benefits without an MES, but it is highly likely that they will not be accomplished at anywhere near the same rate as when an MES is used. Any quality or process issue that is not captured will continue to cost the company money (a direct load on profit) for as long as the issue exists. Although a company can run a CI initiative or plan production without an MES, the ease of pinpointing the root cause of an issue and implementing a stable solution will likely be orders of magnitude faster with an MES, provided that it is selected, configured, and supported properly. The remainder of this chapter provides guidance on doing just that.

11.3 What Are Your Business and Manufacturing Models?

Many of the decisions that must be made with an MES initiative depend greatly on the type of business that a company engages in and the different manufacturing models that exist within the enterprise (discussed in Chapter 10, "Differences in Manufacturing Models and Their Impact on MES"). These key points will drive many decisions regarding which MES application to acquire and the interfacing requirements needed, which must include visibility to the company's growth plans. Whether a company intends to grow vertically (incorporating more of the supply chain into a single enterprise) or horizontally (providing specific subproducts or services to many customers as a part of a diverse supply chain), understanding an enterprise's business and manufacturing models will affect the decisions made going into an MES initiative.

11.3.1 Horizontal versus Vertical Manufacturing Alignment

In companies that are vertically aligned, as in Figure 11-1, subassemblies and data from as-built records or device history reports (DHRs) must be available to other plants, permitting products to continue the manufacturing process.

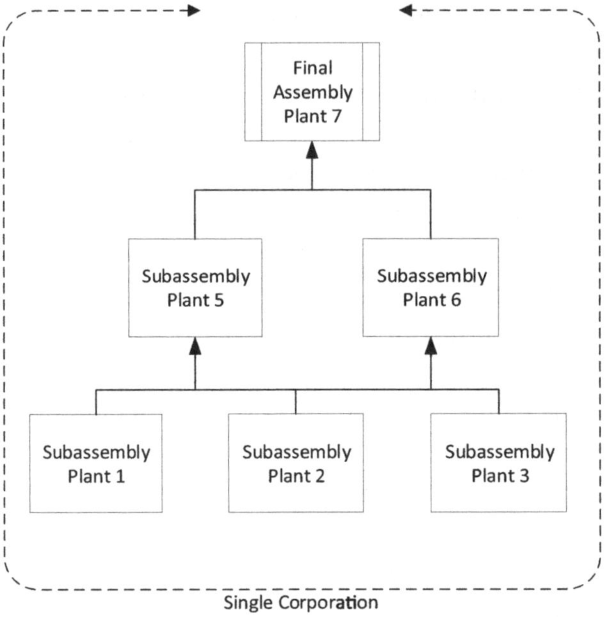

Figure 11-1. Vertical alignment of plants.

In most cases, each plant will be in the same enterprise infrastructure and, as a result, will be able to access the full DHR at the next level of manufacturing. For these types of environments, the basic concept of service-oriented application (SOA) interfacing will probably be sufficient. Also, in this environment, determining the root cause of a problem at one level may require access to the build records at a different plant level within the manufacturing enterprise. Having free access to the data of different levels of the build process can make solution development much more effective. It can also provide an opportunity for plants to lower the costs of IT services by sharing installations of the MES application. There are configurations for installing an MES application that may result in a lower cost to support and provide catastrophic failover capability. These options depend on things such as the relative (cyber) distance between plants and differences in manufacturing models.

When a company is horizontally aligned (supplying similar but limited subassemblies to diverse customer companies or original equipment manufacturers (OEM)), as in Figure 11-2, the perspective of data movement changes considerably. In these

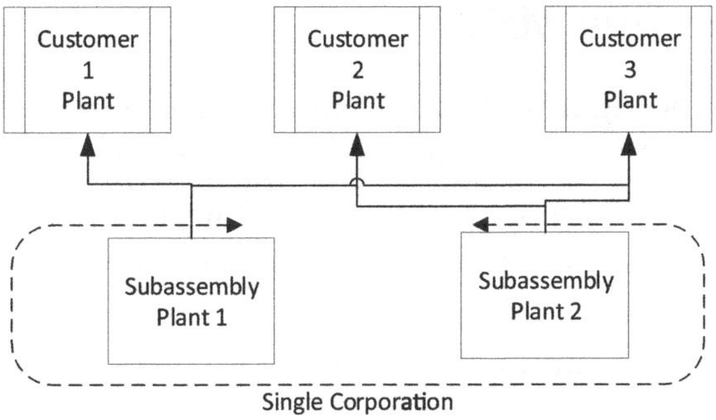

Figure 11-2. Horizontally aligned plants.

business models (e.g., the auto industry), the horizontally aligned company has limited visibility of the end product (the specifications of the specific end product the OEM is making); therefore, the build history is limited to the subassembly that is being made.

Depending on the contract with the customer, communication of the build history in an advanced ship notice (ASN), for example, may be limited to the serial numbers (or batch IDs) included in a shipment or may require a more detailed report. The details and delivery requirements of the ASN can have a significant effect on the interface requirements of the enterprise as a whole. In this business model, the manufacturing and quality engineers will have limited access to the data required to determine the root cause of an issue and, therefore, the solution to be implemented. If the company has multiple plants that produce similar products, it would be beneficial for the plants to share insights and information, but only if they have similar manufacturing models and MES configurations supporting those models.

As noted in Chapter 9, the company's growth plans (organic versus acquisition) and the plan for structuring the system interfaces with alignment to ISA-95 should also be a part of that decision-making. Some application vendors provide a manufacturing solution that supports the entire manufacturing data stream within one system (that handles both enterprise resource planning—ERP and MES requirements and ignores the ISA-95 framework). Provided the company maintains the same business model for many years and does not need to change the enterprise applications, this structure of IT support for manufacturing will be acceptable. However, it will come at the price of locking the company into that application set because it becomes too costly to change both the ERP and MES levels of functionality with different applications. Using a single monolithic application to support both the ERP and MES level of functionality would not be following best practices and is not recommended.

11.4 Who Will Use the MES?

When planning and funding a project for enterprise-level applications (ERP, product lifecycle management—PLM, etc.), consider that a wide range of departments and groups of people use the system and interact with it daily. As a result, it makes sense that the department common to all (i.e., the information technology—IT department) manages and funds the implementation. This is one of the reasons the IT department frequently reports to Finance and why IT is usually the project owner.

However, with an MES implementation, the primary users (and frequently the only users) are Operations staff, and the primary benefactor is also Operations. There is a general rule that says, "he who holds the budget controls the project." For this reason, it is highly recommended that an MES implementation be managed (and funded) by Operations. Some functions related to an MES implementation will affect and therefore involve IT (e.g., server and networking maintenance, databases, operating system compatibility, and application installation and upgrades), and if the implementation includes customization (which it probably will), the project may require code development skills. But beyond these points, an MES (and the management of the initiative's budget) should be an Operations responsibility.

Many MES implementations have been controlled as IT projects, which has led to poor decision-making as it relates to the optimal support of operational requirements. However, if a company has a branch of IT that is highly operations-centric, problems can be avoided. The best practice for managing systems that support operations (from production planning to plant automation) is to control them under a common scope of management. For this reason, some companies have created an IT/controls engineering group to manage their IT and operational technology (OT) capabilities (including automated equipment, programmable logic controller and supervisory control and data acquisition—SCADA systems, MES, and parts of ERP) that reports (at minimum a dotted line) to the head of Operations, separate from the remaining enterprise-level systems (networking, server support, email, web services, etc.). Determining who will own the budget, who will control the project management, and the reporting structure of the support staff must be considered in the decision-making process for an MES implementation.

11.5 Implementing the MES

When this book was published, there were more than 400 MES applications on the market (not including custom internally developed systems), each with its strengths and weaknesses for the particular market it was designed to operate in. Determining which MES application is best for a company depends on several factors. As stated earlier, each production environment has its own products that are being manufactured, its own set of equipment to do the manufacturing, its own people (including skills available and culture), and its own enterprise systems that must be interfaced to. As a result, each MES implementation will have a considerable level of uniqueness. Knowing how an MES will change operations and the impact any application and system configurations will have on operations is critical. The concepts presented in this section should be reviewed by a company's steering committee prior to selecting the MES application and the implementation team.

11.5.1 Matching the Manufacturing Model

In Chapter 10, we discussed several manufacturing models and some of the functionality needed to support these models. As part of the lead-up to selecting an MES, the company must understand the manufacturing models that best describe their plants in order to understand which MES applications would best support them.

When reviewing an MES application, it is best to be cautious and to ensure that the design is a proper fit for the manufacturing model. Many MES vendors will gladly add extensions to their applications at the expense of the client and then use the new function as an advertised new capability.

When a company has multiple manufacturing models to support, the decision becomes a bit more complicated. In most cases, an MES application that supports a diverse number of *discrete* industries will be easier to customize to support *process* and *batch* industries than it will be to customize a process or batch application to support discrete industries. Depending on the extent that a company is diversified, it may be better to use multiple MES applications.

The decision to use multiple MES can be further supported if the application interface is using a properly designed SOA function at the enterprise level.

One criterion to use in the decision-making process is to look at if/how the different products at the plant level roll up in product structure and if there is a need for one plant to send DHRs to another plant to continue the manufacturing process toward the end product, as covered in Section 11.3.1, "Horizontal versus Vertical Manufacturing Alignment."

11.5.2 System Types

As part of the scope of decision-making, a company must decide whether to purchase a commercially developed system (commercial off-the-shelf—COTS) or to develop its own MES internally. In this section, we will review the pros and cons of each of these systems and introduce additional alternatives.

COTS Applications

There are several benefits of using a COTS system that should be evaluated in relation to the long-term growth of the company. Within the time that MES applications have been on the market (since the late 1990s), the knowledge base for developing COTS applications has grown. Additionally, over the last couple of decades, many global manufacturing companies and organizations (e.g., the International Society of Automation—ISA and Manufacturing Enterprise Solutions Association—MESA) have pooled knowledge and have been developing and implementing best practices for manufacturing operations management (MOM). The COTS MES vendors have been working alongside these organizations and have been developing functionality to support manufacturing based on those best practices.

Many smaller manufacturing companies, alternatively, tend to have unique business and manufacturing processes that may not be supported by larger applications. Additionally, there will be specific process differences between industry sectors regardless of best practices. As a result, some commercial MES vendors have developed applications to target specific industries, and some have developed larger, more diverse MES applications to support a wider variety of industries. The variety may come at the cost of support in any one industry. The more targeted applications may provide deeper industry-level support for specific manufacturing models but may make it difficult to expand into other manufacturing models.

Regardless of the industry, as smaller companies mature in business and process management according to process capability maturity model (PCMM) definitions, there is a strong tendency to migrate toward industry best practices to keep costs down and remain competitive. How quickly smaller companies expect to migrate to these best practices will be one of the factors that determine which MES is the best fit. In the investigation phase of an MES initiative, companies should have an independent process-level analysis performed (to understand their PCMM level and their manufacturing model) to ensure a better match for MES support.

For most manufacturing companies (large or small), there is a tremendous benefit to reviewing the ISA-95 standard (particularly ANSI/ISA-95.00.03-2013 Part 3[1]) as part of their ongoing process management (to achieve higher levels within the PCMM framework) and to use this part of the standard to highlight needed processes that have yet to be defined. Adopting the activity model as a guide to defining required processes can help to better match a company's requirements to a specific MES.

---

1. ANSI/ISA-95.00.03-2013, *Enterprise-Control System Integration – Part 3: Activity Models of Manufacturing Operations Management* (Durham, NC: ISA [International Society of Automation]).

It is important to recognize that even if a process within the company is not explicitly defined, it does not mean that an activity within the standard is not being performed. It could just mean that there are multiple variations of that process (and variations in outcome) that are uncontrolled and, therefore, must be reconciled at the time of implementation.

One aspect of COTS that requires investigation is how the MES application will be hosted: internally on the company's internal infrastructure, via the cloud, or on a hybrid system using cloud and fog or edge technology. The differences were discussed in Chapter 9.

To determine the hosting strategy, it is essential to review the characteristics of the enterprise and the system response time required by the production floor. For example, if a production line has a takt time of less than 1 minute, a system that takes up to 3 to 5 seconds to respond to each transaction can quickly use up as much as 30% of the takt time to complete a single production unit at an operation. This kind of delay would be a complete disaster for production planning.

Using an internal hosting strategy (see Figure 11-3), once the server and database sizing is established for the MES requirements, the resulting configuration is dedicated to the

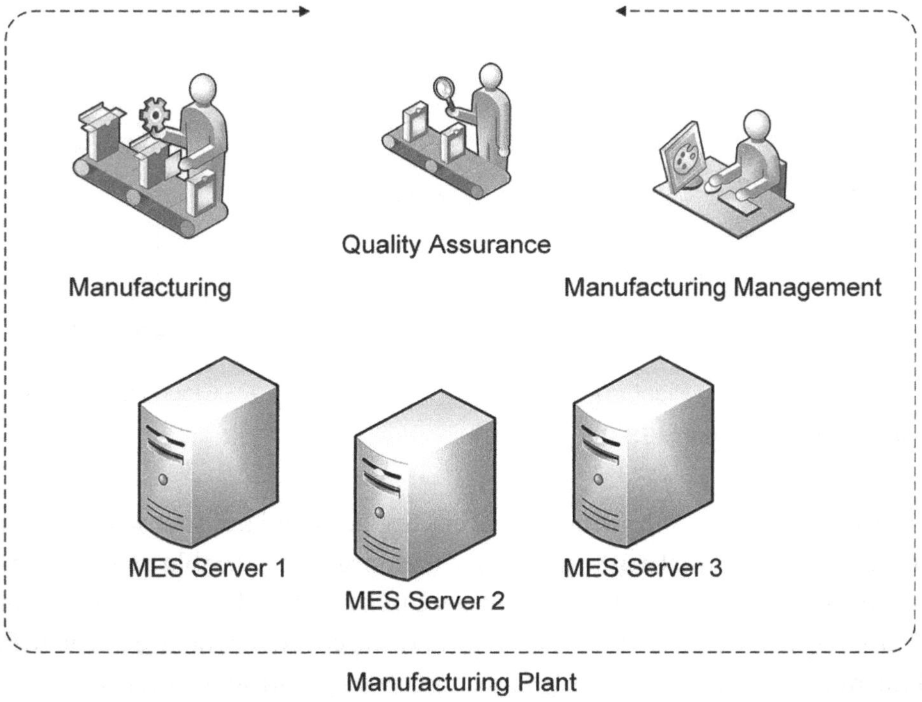

Figure 11-3. Internally hosting the MES.

MES application and fixed to the set parameters (even when using a virtual machine configuration). The cost of acquiring these resources is also determined upfront as a CapEx (capital expenditure) requirement.

Depending on the company's market, there may be wide variations in computing resource usage based on seasonal shifts in production. As a result, computing resources may be underutilized or overburdened (and therefore slower) as demands on production change for whatever reason. If a company has relatively consistent rates of production and growth, the number of computing resources can be monitored and adjusted as required. When there are issues as a result of resource availability, the easy solution is to simply acquire more computing power or hard drives for the database. With this control, the response time of the MES application is maintained at acceptable levels for speed. In highly automated production environments, real-time response is very important. In more manual production environments, there is more leniency with system response but only to a limited extent.

Cloud computing may be a better solution (see Figure 11-4) for companies that have wide swings in production rates as a result of seasonality, an introduction to new markets, or other reasons for rapid change in production volume.

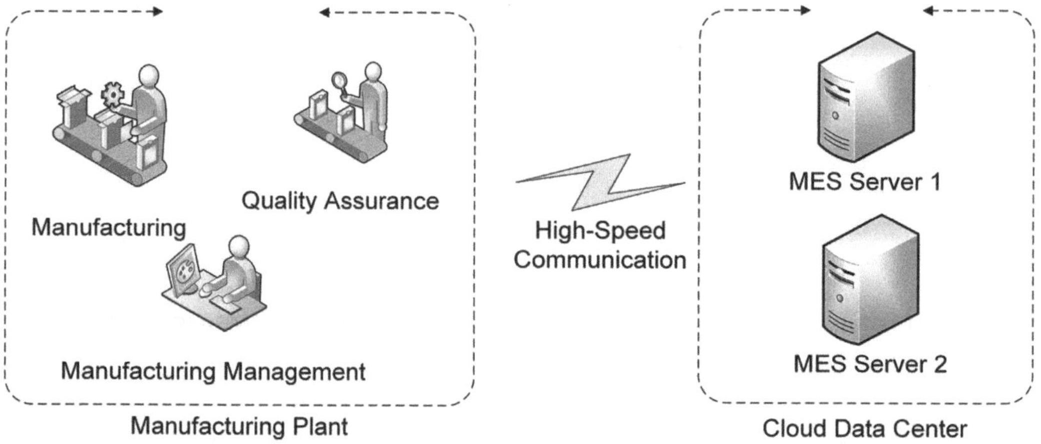

Figure 11-4. Cloud hosting the MES.

In this environment, the computing resources are adjusted automatically by the cloud provider, and the manufacturing company is charged a variable rate depending on the resources used. The computing resources can be covered under OpEx (operational expenditure) requirements and therefore spread the cost over the life of the MES application. This may affect the cash flow of a company's operating expenses at a time when money is tight, but computing resources are needed to handle sudden market growth.

An additional issue with cloud environments is that the response time of the application can vary greatly because of forces external to the company. Depending on the location of the cloud computing center, there is also the possibility that the computing center will go down (e.g., because of a power outage) and not the manufacturing plant. This would, of course, leave the manufacturing plant crippled. Good cloud providers have hot backup systems that would probably be more than capable of maintaining service, making this a likely nonissue.

A fairly recent advance in computing is the use of *edge* or *fog* technology (see Figure 11-5). This would provide the manufacturing plant with the real-time computing speed of on-site hosting and the variable demand capacity of cloud hosting. In the MES industry, this technology is still new, and many applications may not support this hybrid environment.

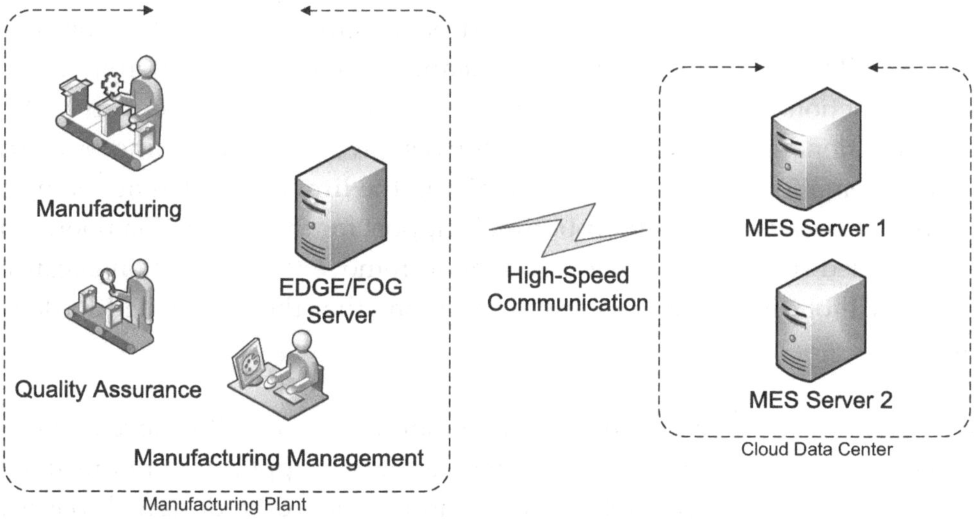

Figure 11-5. Cloud hosting with edge/fog.

An important aspect of COTS applications is customizing the application's functionality. Regardless of how long an MES application has been on the market, there will always be gaps in the support of processes and procedures that a company requires for MES core functionality. As stated earlier, each company has its own products, equipment, and people that contribute to unexpected differences in functional requirements and create the need for customization. Many COTS applications have been designed to handle customizations effectively without having to rewrite a lot of code.

With cloud computing systems, the ability to customize the application will be *very* limited. In cloud systems, the provider only maintains a single version of the application to be used by all clients. If the application provider were to try to maintain multiple versions of the same application, a model such as private hosting referred to as *software as a service* (SaaS), they would spend too much money trying to support the different versions. As a result, most cloud applications will not support customizations or multiple active versions. Any new functionality requested by a client must be reviewed for the product roadmap as a core functionality. This process will result in much longer lead times for delivery of the required functionality and may not be developed in a manner that is in the best interest of any particular client.

In some cases, the application provider will support the application in a *private cloud*. In this model, the company maintains its own internal cloud environment and the applications that reside in the private cloud. This configuration provides the flexible resource handling of cloud systems, however, at the cost of supporting all applications privately.

The ability to handle customizations should be a significant consideration in selecting both the MES and the implementation team. Because of the differences between any two production floors, it would be unwise to expect to implement an MES using only core functionality. In the past, IT executives have placed this kind of restriction on an implementation team without recognizing the limitations that are being set on manufacturing because of less-than-optimal support for the production floor. This has had the extended result of Operations asking to remove the application because they were faster without it, leading to early perceptions that the MES failed to deliver as expected.

An additional aspect of supporting customization is that once it is developed, a well-designed customization should be *configurable* to handle applicable options in process management. As much as possible, customizations should be designed to *extend* the current MES core functionality. In past implementations, poor designs have been developed to *replace* core MES functionality and have led to companies being locked into particular revisions of an application because the cost of upgrading was too restrictive. This can lead to the loss of support by the application developer and the stagnation of a company's MES capability. When selecting the MES application and the implementation team, the customization strategy should be reviewed to understand the capability of each (the team and the application) to support the company's requirements.

Internally Developed MES

To reduce the cost of implementing some aspects of MES capabilities, many companies try to develop their own MES. Although there is still a considerable cost involved in *homegrown* MES applications, the issue frequently comes down to the difference between OpEx and CapEx. It is frequently much easier for an operations manager to get approval for an increase in OpEx over a period than to get approval for a large CapEx. The question is, will the internal expenditure be a better investment?

When a company engages in developing an internal MES, the system is designed to support the company's current internal business processes, rules, and policies. In addition, the frequent direction of the design team is to reference a single plant as the model of the design. Depending on the quality of the design and development team, it may be difficult to change (or extend) these designs should the company change its business model in the future or when the application is rolled out to other plants.

It should be expected that the MES will cause a change in a company's method of managing the business. Depending on the size and industry of the company, as well as the knowledge of industry best practices, this may or may not be a concern. However, it should be considered as part of the steering committee's decision-making process.

In most situations, the person (or team) who takes on the role of designing and developing the internal MES will do so as part of their *other* full-time responsibilities. Because the design and development of an internal MES is not their primary role, in many cases, the knowledge of MES principles will be limited. This will likely result in lengthy development and testing phases as the team *investigates* solution designs. In addition, understanding and implementing software coding standards will likely be limited, resulting in an application that has used *spaghetti code* to implement the solutions. In past internal implementations, these issues have manifested as frequent *crashes*, causing interruptions to production, and have limited the team's ability to add new functionality. In many situations, the system was developed to support one particular plant, and as additional plants attempted to migrate onto the application, the ability to add additional functionality became a major factor. If the company hires a team that has experience in MES design and development, this may be a trivial concern, but it would likely come at a higher cost. In essence, the team would be developing yet another near-commercial-grade MES to match the 400-plus other MES applications that are already available; therefore, it would likely be cheaper to implement a COTS system.

However, if it is determined after evaluating other COTS capabilities that there is not an application available to support a particular industry, the homegrown MES may be the better solution.

Third-Party Extensions

There are cases when a full-function MES is not needed, as determined during the early adoption phase of an MES or because companies have simple processes. For these cases, some vendors (or consulting companies) have used a development toolbox like an SOA or a human-machine interface software package to develop a suite of extensions that deliver primary MES functionality. These applications, although light on MES function, are inexpensive to license and easier (and faster) to implement. The extensions provide most of the capability for the basic track and trace discussed in Chapter 2, "A Functional Overview of MES," as well as primary reporting requirements. This creates an environment for Operations to *get the feel* of using an MES without a large investment. However, these extensions likely will not migrate easily to a full-function MES environment and likely will need to be replaced at some time.

11.5.3 The Implementation Team

When it comes to acquiring the skills to implement an MES, there are options available. The criteria used to determine the best option depends on the company's requirements and the availability of labor to implement and support the MES environment after it goes live. There are three options for acquiring implementation skills, and the correct solution may be a combination of them. By exploring various implementation options, the company should be able to find a cost-effective solution.

Turnkey Consulting

One typical option for implementation (that has been the norm for decades) is the use of a turnkey implementation team from a consulting firm. The team consists of a project manager, a lead solution architect (specific to MES), business analysts, and possibly multiple configuration and development consultants. Additional support staff may be required for these projects, such as delivery management and quality assurance staff, but these personnel will be required on a limited basis. This implementation team provides all services to implement, test, and commission the MES and is responsible for the project from requirements development to go-live and for a short support period after go-live frequently referred to as *hyper-care*. The knowledge of these teams is predominantly IT-based and can only provide the specified functionality from approved requirements. This places the full responsibility of operational awareness of MES onto the customer.

For these implementations, the company should have a high degree of knowledge in both the business processes that are expected at the time of go-live and the way its business model may change over time. When hiring a consulting firm, the company will have a choice of larger firms or smaller firms. At larger firms, most of the skill set required is provided by internal employees, which enables the firms to have greater potential control of staff availability and quality of workmanship. This will come at a cost, as these firms will also have highly controlled rates for engagement. Smaller firms will have greater flexibility in the rates but will likely need to search for the required skills. This can cause greater variability in the team's ability to implement the MES and in the quality of project delivery.

Because the turnkey consulting team is directly responsible for the implementation meeting their customer's requirements, they must spend a considerable amount of time formalizing the requirements that will be delivered before implementation begins. Any functionality determined after the requirements sign-off will require change order management for the original requirements list.

For companies that already have experience using an MES, as well as knowledge of change management, this method of staffing is effective.

Mentorship Consulting

An emerging alternative to turnkey consulting is mentorship consulting. In this model, the consulting team consists of only one or two consulting members who bring deep knowledge of the MES environment (including management of the MES/Center of Excellence—CoE initiative) and broad technical knowledge of MOM best practices (quality, production management, facilities management, MES data mining, etc.). In addition, they bring knowledge of IT best practices (shop floor networking, server management, programming, etc.) and the processes for interfacing with enterprise systems and production equipment.

In this consulting model, the company specifies a dedicated group of employees to become the company's CoE team, which is made up of a diverse skill set (both operational and IT). These employees perform the actual implementation and establish the long-term support policies for local operations (at the plant level). The mentorship team provides the CoE team with the detailed knowledge required to use an MES for the broad MOM scope of business initiatives and guidance on operational support policies and procedures. The mentorship team also helps the company's Manufacturing and Quality Engineering teams understand how to utilize the MES application, engage in data mining and reporting techniques, and interpret the data in a production context. This guidance helps set up the CoE to maintain the MOM business initiatives long-term and helps the company self-manage the changes of a CI initiative.

This method of implementation will require a more collaborative engagement between the consulting firm and the company but can result in a team that is better prepared to support operations. It will likely result in a longer-term relationship between the company and the consulting firm that may continue long after the application go-live. However, this method of engagement will lower the monthly CapEx of project implementation and spread the cost of the implementation over a longer period, significantly reducing the stress on cash flow. In addition, most of the skill-set development and labor will be covered under OpEx and will be easier to approve.

The company maintains ownership of the implementation, but the mentorship team has ownership of the development and delivery of the knowledge base to both implement and maintain the MES strategy and program over the course of the relationship.

This option should be strongly considered by companies that have many plants that require implementation and/or companies with diverse business models.

Internal Implementation Team

This implementation method is similar to the mentorship consulting practice with the exception that the mentorship team is either hired or developed internally as well. This implementation style would best support a homegrown system as described in Section 11.5.2, "System Types," and may benefit from hiring an external consultant (see "Mentorship Consulting") to jump-start the team.

11.6 Implementing MES Functionality

When implementing an MES application, there are concerns about the cost of implementation, the disruption to the production floor, and getting a return on the investment of implementation. As part of the return, Operations must consider the benefit not only to the production staff directly but also to the production support staff. In terms of the production staff, some of the quickest wins will be defining a stable and repeatable manufacturing process, ensuring that standard operating procedures are available when needed, and ensuring that the procedures are viewed. This provides the production operator with consistent guidance on performing the manufacturing steps for each operation, which helps to prevent errors and implement changes to the process more quickly. Also, there are additional benefits for the support staff (manufacturing and quality engineers, planning and scheduling staff, etc.) in the form of tracking and reporting production orders and process status, as well as nonconformance management, all in real time.

The issue is that before the system can be used reliably for order and process monitoring, the support staff must gain confidence in what the system is telling them, and they must develop an understanding of how to interpret the data and reports that are provided. If the support staff is already familiar with using an MES, the capability of interpreting the data will come quickly. If they are not already familiar with it, this part of using an MES can be confusing and frustrating. When a company has a well-trained CoE, the CoE team can provide support services to prevent this confusion.

As part of the decision-making process on an MES implementation, the steering committee must take into consideration the level of knowledge and experience that is available to use an MES application specifically. Care must be taken to properly understand the in-house knowledge level because failure to take it into consideration can lead to major disruptions in production, resistance to taking advantage of the benefits that an MES implementation offers, and attempts to identify ways around the controls that the system has been put in place to provide.

Depending on the knowledge level of the Operations staff, there are primarily two ways an MES application can be implemented over time. The first is a *big bang* approach, which is to implement and use as much of the functionality as possible right from the start. This method is good when there is already a baseline of knowledge about an MES and when there is already a level of maturity around process management (i.e., approximately Level 3 on a PCMM scale) as well as knowledge of data mining for root cause analysis. The other method of implementation is a *phased* implementation approach that provides MES functionality in limited sets designed to build a foundation of knowledge in one phase before implementing the next phase.

11.6.1 Big Bang Implementation

When a company has considerable experience in using an MES application, it is beneficial to provide Operations with the tools (the MES functionality) to implement process and product improvements quickly. Whether a company is using Lean, Six Sigma, or the Theory of Constraints as its primary continuous improvement methodology, it is important to collect and analyze data, run design-of-experiment tests, and implement changes quickly. The faster the new processes can be determined and stabilized, the faster the next CI issue can be tackled. Making sure that equipment is running well and getting a *heads-up* on potential equipment failures is essential in maintaining key parts of the supply chain, ensuring *on-time delivery*, and achieving better reliability in forecasting and *available to promise*. A good MES implementation will help in these areas. Additionally, providing the tools to the Operations staff quickly is an important outcome of the big bang method of implementation. By using a turnkey implementation team in conjunction with this method, this approach disrupts the production environment for the least amount of time (it will be a major disruption, but the dust will likely settle quickly).

For this implementation method, it is also important that the company has a well-established CoE or staff that has experience in managing a CoE. This team will play an important part in rolling out the application in multiple plants.

The downside of this implementation method is that the company must have considerable visibility of the future process models, including the structure of the data and interfaces that will be needed in those future models, and must be very good at change management. Failure in any of these knowledge points will potentially extend the disruption period for months and can lock the company into a *less-than-optimal* configuration over the long term.

11.6.2 Phased Implementations

When knowledge of an MES is limited within the company, a phased implementation may be better (for details on phased implementations, see Chapter 2). This method enables the company to implement an aspect of the MES, gain an understanding of the reporting and control that the MES provides, and learn how to use the MES to support CI initiatives regardless of the qualitative methodology used. This implementation method will give the CoE (and Operations staff) a chance to *iron out the kinks* at one phase before investing in the next phase of implementation. The CoE will also have the opportunity to establish a well-managed program before getting too deep into the support of additional phases. This methodology is well suited to using the mentorship consulting team (previously discussed in this chapter), which may be considered a downside because of the longer engagement with a consulting team. But with smaller monthly costs, this method will be easier on the cash flow.

This method also enables the CoE to determine operational standards that can be used to build the foundation that the remaining MES functionality will use and limit the risk of long-term errors in configuration.

A downside of this methodology is that it requires a dedicated staff to implement the MES for a long period of time, although the disruptions to operations are in *smaller bite-sized chunks*. It could be said that any good CI initiative will cause disruptions anyway. The Theory of Constraints, Lean, and Six Sigma all make changes a continuous part of operations, and some disruption is, therefore, to be expected (that is why they call it continuous improvement, *right?*).

Whether the big bang or phased implementation is the better method depends on the culture within the company and the costing structure preferred by the steering committee. A decision regarding the implementation method will have an impact on the application selection and a *significant* impact on the structure of the implementation team.

11.6.3 Database Planning

Regardless of the strategy for implementing MES functions, one of the more important decisions that must be made strategically relates to data storage and the type of data inquiries that will be made from each data storage type.

As covered in Chapter 9, all MES will use a form of database schema (from the operational side) that will define how the different data objects (i.e., the blocks of data related to a context) are stored in relation to each other, referred to as a *star schema*. In other words, *the relationship between the material master record and the bill of materials record*, or *the relationship between the operation master record to routing*, and so on must be established. Although this is highly IT-centric information, understanding it from an operations point of view is important.

As stated in Chapter 9, the problem with star schemas is that they are not good at accessing data from the perspective of trends and quality assurance (QA) reports like Pareto charts. For this type of reporting, it is better to have a database structure that has been *flattened* to relate all similar transaction types together. This type of database structure makes it possible to use simpler Structured Query Language (SQL) code to determine trends and produce other QA-type reports much faster. These database schemas are referred to as *data warehouses*. Most MES applications will have some aspect of both an operational database and a data warehouse.

Obviously (if you are an IT person), a strategy must be determined to get the data from one structure to another. This will require implementing an extract, transform, and load (ETL) method as part of the MES database support. The details of coding the ETL methods are beyond the scope of this book and are better left to the IT experts who understand them. Most MES applications are already designed with an ETL method.

From an operations perspective, it will be important to know how often the ETL method should be run. In simple terms, the method will take a *snapshot* of the operational database at a specific period to extract the data. This snapshot then must be *translated* into the data warehouse structure and *loaded* (stored in the data warehouse)—steps that take time and computing power. The timeline needed for the ETL method affects the timeline that the data is available for trend analysis. Because trending reports usually reflect a longer period of time (days to weeks), the impact of a 20- to 30-minute delay may be minimal. There are some manufacturing industries, however, where minutes can have a significant impact on the outcome of a trend report. To understand the impact of ETL methods, Operations must look at the timelines of the data available for analysis and verify that they are acceptable for the scope of trend reporting in their plants. For some trend reporting like overall equipment effectiveness, for example, the need for a *real-time* view of a line's status may require these reports to work out of the operational database, regardless of the complexity. As a result, Operations must understand the impact of data being available from one database to the other and then determine how often to run the ETL methods. Care must be taken to analyze the options and impact from an operational point of view. Poor decisions in this area can either cost the company a great deal of money for computing power that may not be required or make visibility from trend reports too far from real time to be of value. Each company (and maybe even each plant) must review this strategic decision as part of its MES implementation.

Do not confuse trend reporting with the function of trend event detection, where the MES application monitors and highlights the occurrence of manufacturing operations (or products) that have specific data trending in an adverse direction. If Operations is using trend event detection, this function *must* be implemented in the operational database but must also be limited in scope.

Another subject that must be reviewed is *data archiving*. This is the removal of data that is no longer relevant from an operational or data analysis point of view and the storage of that data in some form of off-line record keeping. This is *not* to be confused with data warehousing as discussed earlier (in this chapter and in Chapter 9). In many transactions with an MES, the data related to a transaction must be verified against different criteria before the transaction is performed (this is covered in detail in Chapter 2). As more data records are created as a result of transactions, more records must be reviewed for each transaction validation.

Over a relatively long period (depending on the production environment), this buildup of data records (particularly in the operational database) can have a negative effect on the system response speed. In addition to the data warehousing decision-making, the MES implementation team must review data archiving requirements in relation to operations. One of the primary decision criteria for archiving will be whether data within the system is going to be *statistically significant* or if it will be *redundant* in terms of process characterization. Again, these requirements will be specific to each company and possibly each plant.

In the earlier years of MES implementation, these decisions were frequently *side thoughts* of the implementation ("we must do it at some time") and sometimes resulted in critical problems because the issue was ignored until it was too late.

The recent development of new technology *in-memory databases* has resulted in an impression of data warehousing and archiving being less critical. Although this technology makes data access much faster (and the buildup of data records less recognizable), it does not make this problem *go away*. It simply provides a longer period before the number of records has an observable impact.

A close examination of both data warehousing and archiving requirements must be included as a part of developing an MES strategy and can only be looked at in the context of each company's specific requirements.

11.7 Interfacing with the Company

Regarding interfacing with the company's enterprise systems, it is an unfortunate fact that the impact of implementing an MES is not reviewed as part of the implementation of enterprise applications. In nonmanufacturing companies, the ERP system is the central part of the enterprise infrastructure, and as a result, all other applications have developed interfaces to ERP, including most MES applications. Many implementations have then used the ERP-to-MES interface to handle all MES communications. Although the ERP is central at the enterprise level, the MES application is central at the plant level in the manufacturing environment. The MES application will be the main point of all activity on the production floor for product, production order, and equipment status; it will record all activity that happened during production; and it is the primary system for delivering, monitoring, and controlling all instructions to the production operator and for stabilizing the process. As a result, the MES application must get data to and from many systems, both at the shop floor level (e.g., equipment, programmable logic controllers—PLCs, and supervisory control and data acquisition—SCADA) and at the enterprise level (e.g., from ERP for order and inventory data; from product lifecycle management for component, data processing; from document management systems for the operator instructions; and even from human capital management for training, payroll, and certification management). As stated in Chapter 9, trying to bring all this data through to ERP and then pass the data from ERP to the MES through a single interface is not effective.

When developing an MES strategy, the company must evaluate the current application interfaces that are available and decide what changes are required at the interface level as part of the initial MES implementation project. Designing and developing the interfaces needed at the MES level can be complex. If the company has multiple plants, it is far better to plan for a single SOA layer to handle the interfacing to multiple MES applications than it is to plan the interfacing from multiple MES applications to multiple enterprise applications, all independently from plant to plant.

11.7.1 Interfacing to the Enterprise (Reprise)

Many industries outside of manufacturing that have multiple offices and distributed transactions are already familiar with the function of SOAs. In the manufacturing industry, however, SOAs are rarely used. Developing a good interfacing strategy using an SOA is beneficial not only in reducing the complexity of interfacing enterprise applications but also in making it easier to update or replace any of the individual applications that are already connected.

A well-designed SOA can receive data from any application connected to it, parse that data as required, and pass on the parsed data to any receiving applications (see Figure 11-6). This process ensures that each application receives only the data it needs and that it does not have to handle data blocks that are not relevant to its function.

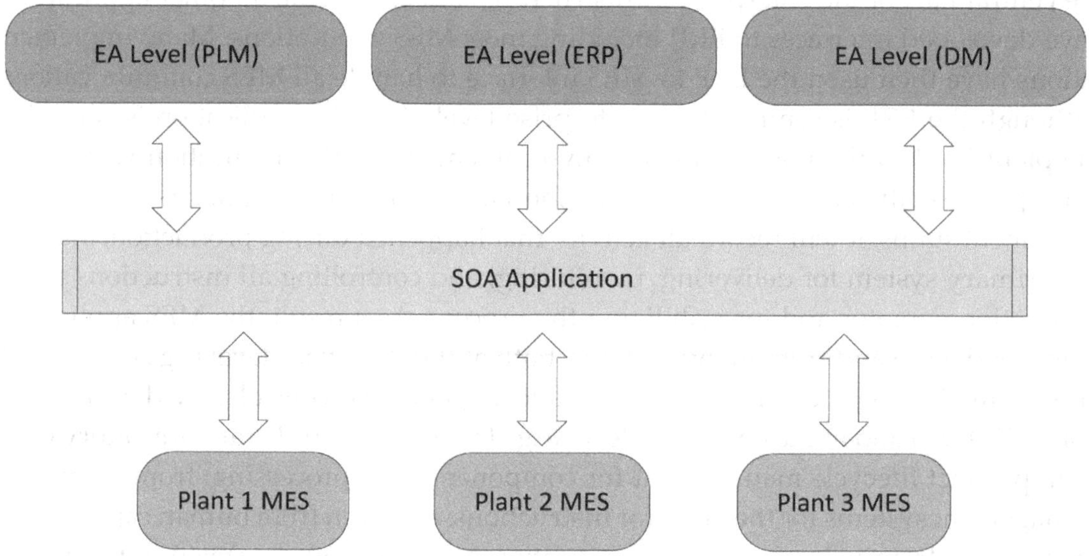

Figure 11-6. Enterprise interfacing using an SOA.

If multiple applications are used for a single function (e.g., multiple ERP systems), as is frequently the case in companies that have grown from acquisitions and mergers, there may be a need for data normalization at the SOA level as part of the parsing process. A company would be better aligned with *enterprise application interface* best practices if it implemented an SOA before an MES initiative. However, interfacing at the enterprise level should be considered in some form as part of the MES strategy.

11.7.2 Interfacing to the Shop Floor (Reprise)

In this age of Industry 4.0, there is a trend to have data made available in many forms and then analyzed to glean insightful information. In the manufacturing environment, the data is collected from various machines and by various means of communication. The goal of the data collection is to provide visibility to changes that are happening at the process control level in an effort to (1) ensure that equipment is operating as expected and will continue to do so, and (2) provide a *heads-up* when equipment parameters are drifting out of control. This visibility is the primary function of predictive analytics as described in Chapter 9.

As part of the planning process for an MES implementation, the company must look at the level of the production floor systems that are actually going to control the process. Most MES applications are designed to control the movement of product (i.e., which production unit will be worked on, at what operation and/or workstation, and when). As a result, if the PLC/SCADA level has also been designed to exercise control, this creates a fundamental conflict in the overall production monitoring system. In many production environments, there is a need to record more activities than are visible to the automated system (this may be argued by controls engineers). Therefore, the MES application is frequently the better choice of the system to maintain control of the process. Whether it is the automation level (ISA-95 Levels 1 and 2) or the control management level (ISA-95 Level 3), the company must plan the implementation and application selection around this requirement.

In Chapter 9, we also discussed the requirement for data collection and data conditioning before the data record is sent to the MES for relational storage. As part of the Industrial Internet of Things (IIoT) initiative, many companies are now looking for equipment that can provide the preconditioned data directly from the equipment. This enables direct interfacing between the equipment layer and the MES layer (or any other application that can use the data). If companies have not established this level of communication (many older models of equipment do not have this capability), they must plan data extraction methods with some other means of communication. This may mean using raw data application program interfaces (APIs) to provide data to and from PLCs or using SCADA systems to access this data directly from sensors.

The primary concern for planning is to ensure that the parameters that can have an influence on the process stability can be collected. The problem is determining which data elements are going to influence the process. As in all things with MES, *it depends* on what part of the process is a concern. This question is complicated by the fact that when equipment is engineered and fabricated, it is not known which data elements will be a concern in the future of manufacturing the many different products that it will likely be making. To account for this uncertainty, companies are now looking to create standardized tags and APIs for most of the operational parameters of the equipment and to allow the querying system to select the required data when it is needed.

Understanding the requirements for interfacing shop floor equipment, whether the equipment is IIoT-capable or not, is an important part of the MES strategy and part of the evaluation requirements for the steering committee.

11.8 Supporting the MES Initiative

After a company has successfully implemented an MES application and gone live with it on a couple of production lines, the next step in the rollout is to ensure that Operations takes full advantage of having the system available. The obvious business processes that are affected by an MES should be examined, including production scheduling, order management, line-side stocking (LSS—inventory), plant maintenance, and quality management. In reality, there are very few business/operations functions within a manufacturing company that will not potentially be affected by (and benefit from) an MES implementation (including some unexpected business functions like finance, customer service, and product engineering). The steering committee should require that the company get support from a consulting firm that can lead the business transformation and highlight the benefits of using an MES company-wide.

As discussed in earlier chapters, the business processes that are affected by an MES (and that can affect the way an MES is used) extend almost end-to-end throughout the company. Setting up a successful initiative is not just about implementing the systems to support the business but also about managing the business processes that the company uses to manage the business as a whole. In this section, we will review activities and business processes that must be established to deliver the full return on investment (ROI) of an MES implementation.

The MES must be supported by upper-level leadership before any work can begin. Their support depends on their understanding of the MES, how the overall organization will be able to support the MES, and ultimately the bottom line: what will be the ROI of an MES implementation?

Initial support based on regulations is fairly easy to get: if the company wants to manufacture a product, it must do so within the confines of the legal requirements. Further support as to the benefit of an MES must be shown in relation to the cost benefit of better order handling and material handling efficiencies, as well as improved production uptime and improved product quality metrics.

Ultimately, the level of maturity the company has in managing its manufacturing capabilities will be an important part of determining whether the implementation of an MES is the correct way to go.

11.8.1 Validating Data

As with all systems, data accuracy is key to validating the information, or as the old saying goes, "garbage in = garbage out." To ensure that the data within the MES application is accurate, an initial effort is needed to verify that data (and therefore the reports) are accurate. This will require auditing the data that is made available within the MES. This is similar to *cycle counting* in inventory management and could be incorporated into any cycle counting initiative. The key will be to not only identify instances of inaccurate data but also determine the root cause of the inaccuracy.

Depending on the quality of the consulting that went into the implementation, the quality of the data collected by the MES will vary. The MES data quality will also have to be reviewed as processes change (both from the manufacturing side and from the business side). As with any inspection process (audits are just a variation of process inspections), as the processes become more stable and reliable, the less a system must be audited (e.g., similar to inventory audits, if a company is achieving 98% accuracy in cycle counting, there is less of a demand in the frequency and/or depth of audits). Maintaining a system with accurate data will be heavily affected by the maturity of process management, again, both for the manufacturing processes that create the data and for the business processes that support production. Consistently and repeatedly entering data and using it to derive information will pay dividends in any CI program.

11.8.2 Automating Process Decision-Making

In the initial stages of MES usage, production operators must make decisions on *the next action to take* (e.g., routing decisions) based on operational outcomes such as nonconformances and test failures. For example, a production unit may be directed to a functional repair workstation or to a workstation for aesthetic repair depending on the type of nonconformance determined at an inspection point in the process. In most cases, the production operator (the quality inspector in this case) will decide on the routing path needed (hopefully guided by policies determined by manufacturing or quality engineers). Once there is a greater understanding of the policies (and the data within the MES), most MES applications are capable of easily programming this decision-making into the production route configuration. In most applications, if the decision requirements can be modeled in data queries at a routing decision point, the actual decision-making can be automated, enabling greater stability in production processing and more repeatable production activity.

As indicated in earlier chapters, the concept of automating decision points can be used for many other activities related to manufacturing, such as inventory replenishment, workstation (or work center) availability, and other functional routing requirements. In the initial stages of utilizing an MES, aspects like line setup will be controlled by supervisors or line leads. This will also include the replenishment of LSS used during the production processes. As validating data accuracy improves and the production processes become more repeatable, it becomes possible to automate LSS replenishments with system calls from the MES to the ERP. This function should also include an acknowledgment of the inventory move after the replenishment has been received by the LSS location.

When implementing automated decision-making, it is best to be cautious and ensure that the process has gained a considerable level of maturity (stability and repeatability) before implementation. Failure to do this can result in significant errors in production management. There is an old saying that "to err is human, to really mess things up you need a computer." However, if automated decision-making was implemented too soon, is the fault in the computer system or in the direction to automate the decisions too early?

11.8.3 Managing MES Application Revisions

Another aspect of the MES that is highly IT-centric is managing version upgrades to the MES application. The issue with some implementations is finding a good time to perform a revision upgrade that is not going to significantly affect production operations. From the perspective of Operations, the typical attitude is "if it ain't broke, don't fix it," and as a result, it is never a good time to perform the upgrade. As it pertains to revision management, the previous attitude can lead to serious problems; a much better outlook is more of a Lean viewpoint: "if it has not been fixed lately, it is probably broken."

In cloud-based implementations, the manufacturing company may not have a choice as to when a version upgrade is going to happen. However, in many cases, if the MES application is designed properly, it should not matter when the upgrade happens from the operations side. This is one of the benefits of *most* cloud systems.

With internally hosted applications, the architecture of the MES implementation can have a dramatic effect on the ease of performing these upgrades, and the steering committee must pay close attention to the ramifications of architecture decisions that relate to revision upgrades.

Regardless of the implementation method (i.e., cloud, internal hosting, or hosting architecture), revision management will be necessary and must be specified as part of the program management policies. Failure in this area will cause confusion for IT and operations support; unexpected variations in upgrading schedules (and therefore impact manufacturing schedules); and unnecessary hours of analysis to determine causes of system errors at particular plants, only to discover that the problem was already solved in a later revision, but that particular plant had simply not upgraded when expected. These types of issues will have a detrimental effect on the reputation of the MES support team and on the confidence in the MES application as a whole.

11.9 Supporting Process Standards

As will be discussed in Chapter 12, "Current Manufacturing Challenges and Trends," process standards are an important part of operations management, not just for the shop floor processes used to make products but also for the business-level processes. Depending on the company's industry and its level of maturity for overall process management, the standards that must be supported may be a result of industry or regulatory standards or of company-defined best practices. When creating the requirements for an MES implementation, it is important to define the standards that will be required for the implementation team to support it and to understand how the system and the implementation team will provide that support.

If the company also has management methodologies like Lean's *Hoshin* process or *balanced scorecards*, the steering committee must understand how the MES implementation will support these requirements. For example, if a manufacturing company has implemented balanced scorecards as their planning and management methodology, most of the tactical scoreboards can be supported via the data from an MES application. The implementation team must understand the details of this requirement.

11.10 Setting up the CoE

The primary concept of the CoE is to develop a centralized knowledge base for all manufacturing facilities to draw from. From the MES point of view, it is frequently misunderstood that the CoE is *not* intended to be an IT knowledge center. Many companies have, in the past, related the CoE to support the IT aspect of the MES (i.e., system architecture, configuration, software development, and analysis of software development requirements). Although these MES aspects are important, limiting the CoE to just the IT aspects of the MES can severely limit the benefits a company gains from an MES.

11.10.1 CoE at the Manufacturing Level

In addition to the IT aspects covered earlier, at the manufacturing level, the CoE is focused on improving the quality and efficiency of processes and working with product engineers to resolve issues with manufacturability.

The activities at this level include:

- Formalizing the CI business program
- Capturing data to facilitate the analysis activities of manufacturing and quality engineers for operational initiatives
- Ensuring that the data captured is being interpreted correctly

When using an MES application, it is important for the manufacturing and quality engineers to understand not only the expected outcome of the changes but also the data required to make decisions on necessary changes. For this level of understanding, the CoE must understand the manufacturing processes, how the data represent these processes, and the business logic within the system that controls the data and product movement. A key part of the CoE is to ensure that the data accurately represents the production floor and that the business logic within the system not only controls the processes as expected but also ensures that the system logic maintains a sense of reality. In previous implementations, it was found that system logic was developed to control situations that were physically not possible on the production floor. This kind of effort is costly and can diminish confidence in the system.

The CoE is not expected to have expertise in all processes and technologies used by Operations; rather, the CoE should have expertise in the general analysis process and in data mining to gain knowledge of the processes and technology used by Operations. The CoE is also expected to understand the manufacturing processes well enough to ensure potential conflicts between various operations and conflicts between processes and systems are considered.

In general, CoE members should be experts in the business management processes and applying that expertise within the company. The CoE is not exclusively an MES function and therefore is not exclusive to the Operations team. However, as Manufacturing is typically more familiar with process management functions, reviewing and managing the processes in and out of the manufacturing floor (both direct interaction and supporting processes to Operations) are good places to start.

Companies that have implemented a CoE successfully have recognized that to really benefit from this business program, it must serve both the Operations side and the IT side of CI. This cross-functional knowledge of manufacturing processes, business process management techniques, and IT is paramount to the success of the CoE team.

11.10.2 CoE at the Business Management Level

In many plants, an initiative to benefit CI (like Lean or Six Sigma) will start in the manufacturing area and will make great strides in lowering the cost of manufacturing. There comes a time, however, when the initiative will start looking at the business processes that support production (again, to improve operations and lower the operating cost of the plant). When the CoE starts to take on the supporting processes (e.g., inventory management, order planning, and facilities maintenance), the initiative starts to reach into business areas that are common to other plants within the company (referred to as the *enterprise-level processes*) and will influence the system configuration and user interaction from all plants. This is where the CoE must have relationships with the business members of all plants and where the creation of company-wide best practices will be developed.

To reduce the analysis effort in creating best practices, many companies will model their processes to the requirements of one plant that is defined as "the one everyone tries to imitate." The issue is that often the variances that occur in other plants are not recognized and not accounted for. With changes to people, equipment, and even the needs of communicating with local suppliers, each plant will have some need to support local requirements. When the CoE is creating best practices for enterprise-level processes, the ability to configure local support will be needed within the solutions created.

In previous chapters, we discussed many of the enterprise-level processes that are supported by an MES and, as a result, will interact with an MES. Management of change to processes and product at the enterprise level requires a strong CoE team, not just at the manufacturing level but also at all levels of business. The growth of the CoE capability should be expected to reach end-to-end within the company.

11.11 Summary

An MES implementation is usually isolated to one plant at a time and is easier to manage by delivering smaller projects at a time. However, as with any journey, there is a need to understand (have a vision of) the destination (the end goal). During the initial stages of an MES implementation, it is important to develop a roadmap of the journey and define a strategy for delivery according to that roadmap. There will, of course, be detours along the way, but as long as the vision is maintained, the destination will be achieved. Taking the time to plan the direction of an MES in the scope of the company as a whole is an important first step. As the old saying goes, "if you fail to plan, you are planning to fail."

12 Current Manufacturing Challenges and Trends

We have reviewed aspects of the business process information, technology influences, and manufacturing types in this book to determine how a manufacturing execution system (MES) has been developed and how it can be configured and operated. So, what is left to review? There are a few current challenges to manufacturing that we must understand in terms of their impact.

This chapter will explore the following challenges and their impact on an MES:

- Aging assets and brownfield maintenance
- The challenges of greenfield plants
- Globalization
- Standards, or lack thereof
- The need for simulation
- Implementing other manufacturing operations management (MOM) concepts
- Trends

12.1 Maintenance of Brownfield Plants and Aging Assets

Brownfield plants are plants that already have a defined manufacturing capability and hardware and are either currently producing products or have recently produced products. Inevitably, these brownfield plants need an overhaul. In many cases, Operations will plan these overhauls to be managed in reasonably small chunks, but often there is a need for a major redo of the entire manufacturing model.

Any changes to the manufacturing environment—be it updating the aging assets by replacing obsolete equipment, improving the manufacturing technology (or operational technology—OT) capability, or taking on a major manufacturing process change—require careful planning and can incur significant costs.

Because the layouts of most plants have been designed for a particular set of processes, the challenge in regard to any brownfield upkeep or improvement activities is to first ensure that all the business processes are properly documented. In many cases, the business processes have morphed over time and are no longer comparable to the original implementation. Preventing this difference between the documented process and the current actual process has been a significant focus of quality initiatives such as Six Sigma, Lean, and the Theory of Constraints. The next challenge is to fully understand what must be replaced or improved and to have a clear plan for implementing the change and ensuring that the company avoids a situation where an extraordinarily large CapEx (capital expenditure) is needed to maintain or improve performance. All these changes must happen without negatively affecting the production schedule, and significant changes usually require training activities regardless of whether the change is in assets or processes. These plans and costs can be significant, and the rationale for the change must be justified by the program, including the need for a time-efficient return on investment (ROI). Otherwise, the activity will be a detriment to the company's financial performance. However, the need to avoid obsolescence should be included in the rationale, not just the cost and ROI. It is well understood that information technology (IT) assets must be replaced over time, and most companies have an active plan to replace aging IT infrastructure. Because most OT assets are designed to operate for a longer period, the aspect of obsolescence must include a review of IT/OT convergence and the impact of interfacing on OT assets.

The investment in the hardware necessary to operate manufacturing is significant—especially in controls, programmable logic controllers (PLCs), networks, devices, and other automated equipment—both in initial investment and in ongoing support. As the technology continues to evolve, these equipment assets start aging rapidly; what is now available through the innovation of Industry 4.0 and the equipment capability of the Industrial Internet of Things (IIoT) (as discussed in Chapter 9), was not available in the early 2000s. At some point, these aging assets will not be able to support some of the changes and abilities offered by the functions newer enterprise solutions provide. Additionally, replacement parts for outdated equipment are often not available or are very expensive. As such, aging equipment assets can become a significant cost when they must be replaced or at least retrofitted. Much of the older controls assets are still functional for the production capability they were built for, and some were made to function for decades; however, they are often either void of communication capability outside the PLC system or their communications capability is quickly becoming obsolete and difficult to retrofit unless upgraded with a custom-designed integration package. The cost and planning to upgrade are not only for the acquisition and installation of new technology but also for training to support the new asset and the removal of the old asset. This includes replacing control assets that have become obsolete on the market and replacing parts that are either very expensive or no longer available. Therefore, the lifecycle of manufacturing assets must also be planned for, which includes the retirement of the assets when they reach the end of their useful functionality.

For some OT upgrades, there are retrofit kits designed to easily add to equipment that will provide at least some communication/data acquisition functionality. These kits can provide relatively inexpensive upgrade capability and extend the usable life of the equipment for several years.

12.2 The Challenge of Greenfield Plants

Greenfield plants are new manufacturing facilities and may even include a new building. These plants have no manufacturing assets or processes set up. In planning the project for manufacturing start-up, the manufacturing processes must first be sketched out and approved for all the plant functions first. This may include integrating the greenfield plant into business processes that already exist from previous plants or defining processes from scratch as the plant is being built. The main challenge is the missed processes that remain undocumented. When a process is missed, the omission may not be identified until late in the project and will probably cause difficulty in including that business process in the support systems (e.g., IT, enterprise systems, or business models). Solutions created to include these undocumented processes will be developed ad hoc, and they will likely not easily fit into the overall process flow. These ad hoc solutions will usually manifest themselves as unplanned costs, as well as schedule delays. In addition, supporting these solutions (as stated in previous chapters) can be difficult in the long term.

Additionally, many (if not all) of the processes that have been defined will not have been verified because the business aspect of the company that the processes support is probably not functioning yet. In this regard, it is important to understand that the company's best practices that are being developed will take time to verify. Reflecting on and using other best practices already defined within the industry will help to minimize this impact. Minimizing variation in processes, developing agile solutions, and recognizing the need for improvement are also required and will go a long way to minimize the overall impact of greenfield plant design and development.

For companies that are engaging in greenfield planning, using the ISA-95 MOM activity model can provide some help. As stated in earlier chapters, the MOM activity model provides a reference to all management activities needed at the plant level. Using the activity model as a template, a company can identify many of the gaps in its current processes and can then highlight the areas where interfaces to the enterprise (Level 4 activities) or shop floor equipment (Level 2 activities) must be created.

12.3 Globalization

In the field of manufacturing, profit margins tend to be relatively small and competition is high. Because of the cost of manufacturing equipment, the cost of entry into the market is also higher than in other industries; this is, of course, why contract manufacturing has been successful (i.e., getting companies that already have the equipment to do the manufacturing for you). But for companies that get the formula correct (the right processes, the right type of equipment, and the right product delivery), the market can be very profitable as customers seek out products and services that can be reliably delivered. The key is to ramp up to volume production as quickly as possible. To do this, companies getting into the market must quickly take advantage of the knowledge of best practices already available within the industry. These same companies must also ensure that their growth plans are flexible and, simultaneously, minimize major changes to the enterprise structure. This is the reason that enterprise planning must be a major part of a company's strategy for growth and why many companies struggle.

As information is shared via IT (MES and enterprise systems) and understanding the diverse area of manufacturing continues to drive to deeper details, it becomes essential to develop, implement, and share best practices. This also applies to best practices that are used to support governmental regulations because governing bodies jointly work to help ensure the safety of the manufactured products. These activities are a result of globalization. Another aspect of globalization is the increased availability of skilled resources: as best practices are shared and developed, the knowledge base to improve on best practices becomes dispersed throughout the globe.

One impact of MES on globalization is the improvement of data availability for process and system functionality, which can be applied to both brownfield and greenfield plants and helps to level the competition for western manufacturing companies by significantly improving manufacturing capability. This data availability can be managed through the careful analysis of the data requirements to determine the impact of a change and the required resources to implement the change. These resources also include the timing and costs of these proposed changes. The most tangible impact is the ability to review real-time information, review reports, then manage and at least partially control manufacturing execution remotely. Although some activities cannot be removed from the plant, remote accessibility provides a true location-independent manufacturing support model. Recently, access to remote support for the manufacturing floor has proven to be extremely helpful as companies tried to maintain production environments during the COVID-19 pandemic. The detailed visibility of production activity provided by an MES was instrumental in keeping production plants running while minimizing the exposure of workers not required to be in the plant on a regular basis.

The last impact is on resources. Most MES development can happen anywhere across the globe, but it will eventually require on-site work and support at the plant. Although timing and travel costs can be minimized through remote work, travel will incur additional schedule timing and costs. This causes MES development, implementation, and ongoing support to be affected by geographical, technological, and cultural diversity factors. All this can be managed through standards, which we will cover next.

12.4 Standards, or Lack Thereof

In operations management, one of the primary methods for keeping costs down is to implement standard processes for manufacturing products. Developing and managing standard processes is a key part of every manufacturing operation and quality management methodology. Depending on the particular manufacturing sector, industry standards have been established according to regulations to protect the public; sectors not covered by these regulations have developed and agreed to industry standards to support the ease of commerce. Many companies are discovering this by exploring Industry 4.0 capabilities and the inclusion of Smart Manufacturing concepts. This is only a starting point when it comes to standards.

Not always recognized is the need for standardization for the manufacturing support processes as well as the manufacturing processes themselves. To ensure that the information made available at the production floor level is consistent, the business-level processes that support production must all be standardized. For example, this includes processes such as the release of product documentation to be manufactured, the release of orders or material to the production floor, and the creation of work instructions. Any process that provides information to the production floor must be standardized. In the next couple of sections, we will look at process standardization from both a manufacturing and a business support perspective.

Although each sector may have a good understanding of the standards used within their industry, these standards are not always widely understood by the vendors of MES applications. Following the ISA-95 standards will ensure that common data models and functional requirements are handled. However, care should be taken to ensure that industry standards can be supported by the MES application and to review how they are supported. Many vendors have no issue with developing functionality in data collection and reporting as a customization for a customer as long as the customer pays for it. Any function of an MES that is designed to support an industry standard should be developed as part of the core MES capability.

To obtain clear information on any industry standard, one should consult any of the various nonprofit organizations that support and document industry standards. While an exhaustive list of these organizations is outside the scope of this book, organizations such as the Society of Automotive Engineers, the Society of Manufacturing Engineers, the International Society of Automation (ISA), and the Manufacturing Enterprise Solutions Association (MESA) are reliable sources of documentation on standards.

On a side note, as part of the capabilities of the IIoT, information is also available on the standard use and maintenance of equipment, and suppliers are now able to provide feedback to companies in regard to the effectiveness of equipment maintenance and how a company stacks up compared to others for key performance indicators such as mean time before failure.

As we discuss standards, note that any standard that affects manufacturing or an MES will likely affect logical functionality, system architecture, data interfacing, storage, and user interfaces. The complexity and diversity of manufacturing lead to many different areas in which support for industry standards can provide value. The last area of standards is at the company level, and these standards are developed and supported solely by the company.

12.4.1 Company Manufacturing Process Standards

In any discussion on operations management, there are always points made regarding standard procedures for the production floor that are usually documented in standard operating procedures (SOPs). The procedures, of course, are unique to the actual steps needed to complete the manufacturing process for each product. However, there is direction that can be taken in identifying which operations (and steps within the operation) are needed and in presenting that information to the production operator. The key tasks of the production operator are to make products and to help improve the process of making products. Any time a production operator is distracted from one of these two functions, their effectiveness is diminished and the rate of production and/or the quality of the product are affected (even if the effects are only marginal). Aspects of SOPs that can be standardized are terminology, presentation of information, and collection of data. Creating standards at the SOP level in these three areas will significantly help the training and effectiveness of a production operator.

Many companies encourage and even require manufacturing engineers to produce these SOPs. Often, the Operations management department has a standard format for creating an SOP, but in some cases, each manufacturing engineer will create an SOP under his or her own format with his or her own set of dialogue and terminology to instruct the production operator on how to do the particular operation. In these situations, each manufacturing or quality engineer will have a different mode of presenting the information to the operator. Unfortunately, this makes it more difficult for production operators to learn the operation and to cross-train production operators for a more agile production floor. It also opens up the possibility of misinterpreting instructions as terminology changes slightly from one SOP to another.

The key directive for any instructions to production operators (either SOPs or training sessions) is to provide the information from the point of view of the production operator and in the context of what the operator is doing at that time. This helps ensure that operations are easily trainable, cross-trainable, and understood. It also helps ensure that mistakes are not made. All SOPs should also include actions that are put in place as poka-yoke methods (discussed in Chapter 5), and training should include why the poka-yoke method has been put in place. The last point is that SOPs must include instructions on health and safety concerns. Managing safety instructions consistently (either by training or SOPs) goes a long way to ensuring production operators' safety.

Creating the content of SOPs is not normally in the scope of the MES. For this function, companies should rely on developing templates for content creation and document management (DM) applications to manage revising and publishing the SOPs. After the content has been created, interfacing the DM to the MES application will be much more streamlined and error-free from an interface perspective and will present information to the production operator in a recognized format and sequence that directly relates to the work being performed. In recent years, there has been considerable improvement in web design tools that enable SOPs to be updated dynamically with data directly related to the status of the production floor at that moment. For companies that have complex products, this capability, when interfaced with the MES database, may be of considerable value for some production environments.

Another part of the production operator's activity relates to the tools and equipment being used. Standardizing particular types and even brands of tools and equipment will also simplify the production operator's job and the design of any interaction with the MES.

12.4.2 Company Business Process Standards

Companies take on initiatives such as Six Sigma, Lean, or the Theory of Constraints with the intention of using these tools to build a set of company business standards. When these initiatives are properly supported by senior management, they provide a methodology to step up operations from Level 1 of a normal process capability maturity model (PCMM) rating to sometimes Level 3, where standards are documented, maintained, and followed at the plant level. On the manufacturing floor, an MES is an excellent tool to help build and maintain these initiatives.

Research by organizations related to business process management has found that very few manufacturing companies ever extend to PCMM Levels 4 or 5, where standards are used across multiple plants and are managed by a group representing the knowledge across these plants, similar to a corporate-level Center of Excellence (CoE).

Companies that are currently at Level 1 or 2 in a PCMM evaluation should seek guidance to develop a business program that will help to implement and maintain standards at the company level. When a plant reaches Level 3 in a PCMM evaluation, it can be assumed that the business program is in place to maintain the standards, and the program can then be extended to other plants. At this time, companies can also define the quantitative measures of process improvement and work on optimizing to achieve a PCMM Level 4 or higher.

An effectively running CoE can increase the rate at which processes are tested and improved, help to rapidly stabilize processes for introducing products to new plants, and be a major source of information into industrywide changes. Companies that neglect changes in the industry are at risk of being caught by surprise when there is a significant technology change, and a CoE can help keep manufacturing plants on the leading edge of change or at least prevent them from being left behind. It would also be of value to have the members of a company's CoE engage with organizations like MESA and ISA to help stay abreast of these industrywide changes.

Another challenge in the industry is the activity relating to Industry 4.0 and the IIoT. As discussed in Chapter 9, new solutions emerging in this area provide the challenges of new devices and technological solutions in an area where standards are not mature. Careful evaluation of the benefits of these solutions, especially compared with IT operations, is needed to ensure smooth operations.

12.4.3 MES Architecture Standards

There are as many standards as there are separate manufacturing industries in the general manufacturing sector. Each provides guidance on quality and process management in the context specific to its industry. From the perspective of enterprise architecture for manufacturing, there is a growing consensus that the ISA standards (ISA-95) for defining processes and, therefore, integration of applications used in manufacturing IT that support these processes are very effective. These standards provide guidance on the demarcation of processing (from an enterprise IT view), data structures, and management activity to be supported. For a detailed understanding of the ISA-95 standards, see Chapter 2.

In the area of industry standards, there are two main issues:

1. Although there are standards in place for MES-level data and communication, such as MESA's Business to Manufacturing Markup Language (B2MML), they have been slow to permeate and be implemented in the global industry.
2. These standards are still developing.

As gaps in the standards are uncovered through investigation by diverse industry groups, these standards are updated and new standards are created. As these industry-level standards continue to evolve, they can, in conjunction with coordinated company standards, help improve the agility of many manufacturing operations and the speed at which companies ramp up new and improved processes. The availability of standards enables small manufacturers to implement processes that take advantage of the research that has already been done and will likely cover a large proportion of their requirements.

However, for large manufacturers with several plants, more effort will be required to combine industry and company standards to fulfill all the company's requirements. In circumstances where companies have grown through acquisition, managing standards becomes even more complicated as the acquiring company and the acquiree often have their own sets of standards. This is where effort may be required to reconcile the differences.

However, the effort required to ratify these standards is well worth it. A lack of standards leads to potential variance in interface requirements, which in turn leads to additional costs that provide no return to the product or the company. In addition, larger companies that do not utilize standards are potentially causing harm to their customer satisfaction ratings. From the customer's perspective, each plant essentially appears to be an independent supplier. If the supplier provides the same product from multiple plants, this issue becomes very notable. If a company resolves a problem at one of its plants, it should share the solution with its other plants. When this does not happen, the customer has a right to be unhappy because it places a greater burden on the customer's quality assurance team.

In conclusion, if there are standards, they must be followed, and management activity must be defined to support their ongoing improvement. If standards are lacking, then an effort to create them must be undertaken.

12.5 The Need for Simulation

Any company that has not implemented simulation software must manually evaluate manufacturing details and assumptions. This means that the systems, workstations, and process flows are all manually validated before production begins. This process has been part of the Lean manufacturing concept for years. One example is an analysis of a workstation layout, and the flow of effort by an operator to complete the operation that would have to be observed and timed manually. If a change was needed, such as changing the placement of a subpart rack to shave a few seconds off operator time or to minimize operator reach for health and safety reasons, it may have iterations of physical changes evaluated to determine the optimal setup. Another example is the effort to ensure that any tooling and robotic clearances are correct in relation to the part coming down a conveyor. This has to be verified by building a frame representing the spatial needs of the part on the conveyor and watching to determine whether the robot or tooling collides with the frame. If there is a collision, the robot must be either physically moved or, at a minimum, reprogrammed.

When simulation software is used in conjunction with MES data (or when an MES is used to expand on digital twinning for IIoT initiatives), not only can the clearances be verified but operator interaction and product/process flow can also be verified, including takt time and throughput. Not only is this important to manufacturing but it is also important because it helps define workstation specifics before they are configured in an MES. In this way, an MES and simulation work hand-in-hand to develop and validate process characterization. Any changes in physical configuration will incur additional time and costs. If these items can be validated via simulation before the hardware is implemented, it will save time and costs.

12.6 Implementation of Other MOM Concepts

As detailed earlier in Chapter 1, an MES is only part of the entire MOM function. Enterprise resource planning (ERP), product lifecycle management (PLM) processes, and other manufacturing processes are all included. Because an MES is part of the support for MOM, there are other considerations to plan for as well. However, being at the heart of MOM operations, the MES does have the potential to play a major role and should be considered in the full scope of MOM planning. In executing manufacturing, the MES also interfaces with ERP for all aspects of production scheduling purposes, as well as materials coordination through the entire supply chain, including planning plant storage space for all production parts. Regarding the PLM, the MES must interface with all product-specific data, including the key process parameters, traceability data, and interfaces with work instructions. Refer to Chapter 10 for more details.

A good reference for the full scope of MOM is to investigate the "four pillars of MOM," which include production management, quality management, inventory management, and production maintenance management. These four pillars are covered extensively in the activities of the ISA-95 standard.

All these activities must be documented, understood, and planned for in an MES.

12.7 Trends

A few upcoming trends will have an impact on how MES and MOM solutions are developed, deployed, and supported. These trends will tie into Industry 4.0 and IIoT activities.

Some people tend to see anything regarding the IIoT as disruptive to an MES. This is far from the truth; as the IIoT continues to evolve, it represents the next generation of MES capabilities, whether they are newer and better MES functions or improvements to the data flows and architecture. As with many newer solutions, these newer MES solutions may require a direct replacement of existing systems and hardware, meaning it would require that brownfield plants remove and replace their current applications. For this level of change, it would be better supported initially in greenfield plants, migrating to brownfield locations over time.

One of the most interesting trends is the concept of block chain technology and/or publish/subscribe technology over the Internet (or intranet), as discussed in Chapter 9. Here, the data flow between the different hardware devices works on a data bus model, similar to the produce/consume data model that some PLCs utilize. This would minimize the need for separate communication protocols and simplify the data configurations by subscribing to a data broker instead of configuring the details of the data source as defined by the communication protocol (device type, network address, data address, data type, etc.). There is not much currently available on the market for these solutions, but this should change in the next few years. Not only can IIoT simplify an MES but it can also simplify MOM interactions, including ERP and the PLM.

Another trend, also detailed in Chapter 9, is the ongoing evolution of the autonomous (or agent-based) MES, in which interactions between the various clients of the shop floor (nodes) communicate peer-to-peer, and nodes within the system are smart enough to practically configure themselves. While this is currently in the concept stage, there is interest in evolving it into a viable solution. One of the challenges will be establishing a standard communication protocol similar to those prescribed by the World Wide Web Consortium (W3C) for the Internet and implementing that standard into a usable manufacturing service bus to manage the release of services and the security configuration of nodes.

One more trend worth considering is the ability to start evolving the data model for order scheduling into a set of standards. Currently, each original equipment manufacturer (OEM), and sometimes even different product programs within each OEM, has unique data models for the order scheduler based on the specifics of *top-level* parts. This requires each system that uses the order schedule, both ERP and the MES, to create a unique data translator to repackage the data model from one system to another. Development and support costs will decrease if there is a standard data model to follow. MESA's B2MML is designed to help address this issue.

The manufacturing community is realizing there is a common approach to an MES, and this will affect how MES solutions will be marketed. Currently, MES is marketed to promote capabilities based on the lack of clarity in understanding what an MES is and what an MES is not. Some MES are marketed for niche functions that are so specific they represent only a small proportion of the total MES capabilities. MES solutions are also being marketed as "cloud capable" without providing much detail as to how their cloud-specific solution is an advantage. In this area, greater understanding at the client level will help drive greater clarity and better marketing, especially because MES solutions must likely leverage both cloud and local processing by using fog or edge-based solutions. As these solutions become clearer, their marketing should improve as well.

As a result of increasing interest in the value MES provides, another trend is to further automate the processes in manufacturing—not only plant-floor manufacturing processes but also management processes at the ISA-95 Level 3 and Level 4 business processes. Discussion on this topic began in 2020 because of the pandemic and plants' inability to run production when human resources were not available. MES can provide the platform to support this. However, each process must be fully investigated and understood before attempting to automate the process so that there will be no negative impact on the plant's current ability to deliver the product on time.

12.8 Summary

In this chapter, we reviewed many of the technological and business factors that will be major influences in the planning and development of an MES solution. Taking these factors into consideration will save time, money, and headaches down the road.

13 Bringing It All Together and Selecting a Solution

We have discussed many concepts related to the different aspects of operations management, such as continuous improvement, capacity planning, facilities management, and others. We have shown different ways that a manufacturing execution system (MES) can help with each of these areas of manufacturing operations management (MOM).

We have also covered several process and quality management methodologies and shown ways in which MES can help with them.

We have presented the information technology (IT) side of an MES in a manner that is relatable to an operations point of view and provided guidance on how to use the data from an MES to help many of the enterprise-level business processes perform better.

In this chapter, we will summarize some of the concepts from other chapters and relate them to the bigger picture of an MES program. Additionally, we will present some final feedback on implementing MES in general.

13.1 Supporting MOM

Within the activities of MOM that we have discussed so far, two aspects of operations management are important.

The first aspect is understanding the actual capacity and the requirements for estimating the support of customer demand and then ensuring that the actual delivery meets the actual demand. The issue MOM managers must deal with is that depending on the situation, there are several things that can cause variance in the ability to deliver, and unexpected variances of significant magnitude can be highly detrimental to the planning and management of delivering on that demand. These variances are the pain points that frustrate MOM managers and leaders and cause them to lose sleep (and sometimes their jobs).

The other aspect of operations management that is important in MOM is locating the problems that cause variances within the processes (i.e., manufacturing and business processes) that are a plant's pain points and determining just how big of a problem these pain points really are.

Sometimes these pain points are caused by a high volume of small variances as a result of manufacturing operator errors, and although each problem may not cause considerable concern in and of itself, the volume of issues can still result in considerable costs to operations. These issues are collectively categorized as *common cause* quality issues, and they are problematic not only because of the cost associated with them but also because they create a lot of "noise" for quality management programs and make it difficult to analyze and quantify major issues. In this case, MOM leaders and managers require assistance to stabilize these operational activities and invoke a serious process management program. In these situations, using an MES to ensure repeatability and consistency in a production operator's performance is very helpful. This is also part of the focus of initiatives such as Lean and Six Sigma. This activity is covered in detail in Chapter 5 and Chapter 7.

When the volume of variation has been reduced, Operations can turn their focus to the issues that fall under special cause variations. These are issues that have specific causes that may or may not be known. After the common cause variation (the noise) has been reduced, it becomes easier to characterize processes, analyze the root causes, and take action to prevent these special cause issues. This is also covered in Chapter 7.

The staff's availability to identify the root causes of these problems and then develop a solution to resolve them is often limited due to time constraints. Because of this, it is important for the MOM managers and leaders to focus continuous improvement activities on resolving the costliest pain points, invest in solutions that provide the "biggest bang for their buck," and ensure that the focus is on the root cause of the problem rather than just a symptom of the problem.

As part of the *gaining the greatest impact* concept introduced in Chapter 5, the Theory of Constraints (TOC) philosophy is that if a solution does not improve the capacity of the constraint and/or improve the delivery rate of good product (thereby improving the ability to sell more product), then MOM improvements are not improving things (or at least the improvements have a minimal effect). Whether you subscribe to the TOC is irrelevant. Making the investment that provides the greatest return is always good management.

In the previous chapters, we showed that an MES, when used as the central point of record for manufacturing, can provide valuable insight into the real characteristics of the plant. These include the business and manufacturing processes (cycle time, yields, and, of course, throughput of the right product to be manufactured), equipment and operator capacity (a product of the process measurements when focused on a specific piece of equipment and product combination), and the operational costs related to cost of goods sold. In the next few sections, we will summarize many of the concepts from previous chapters and highlight those that are important regarding supporting and planning the MES.

13.1.1 Summary of Supporting Processes

When discussing the management of a manufacturing plant, key points must be clear. When it comes to forecasting sales (independent demand) and the target markets, Operations will be able to contribute very little to the accuracy of the demand side estimations and should not try to impose on this part of the business. However, both Operations and Sales should keep an eye on the consumption portion of the forecast for any variance that is happening to both the demand coming in as sales and the delivery to sales. This is important to ensure that (1) Operations has visibility of issues that will affect their capacity to deliver and (2) Sales understands what is contributing to the variance in sales demand in an effort to create better forecasts. There is no greater contributor to errors in manufacturing than to have Operations surprised by an unexpected increase (or decrease) in demand.

The part of the demand/delivery equation that Operations has control over is planning and executing product delivery. As with all planning activities, there is an understanding that once the plan execution starts, it is going to have variances.

However, Operations managers understand that what makes the planning process effective is not just planning what to manufacture but planning for contingencies of known variances. The problem is that there may be too many different things that cause variances for Operations to plan for. So, what can the manager do?

As stated in other chapters, most of the effort of operational analysis initiatives is to (1) collect data as accurately as possible to contribute to planning and (2) engage in initiatives that reduce the cause of variances, thereby controlling the manufacturing process as much as possible and increasing efficiency.

By using an MES application to manage the production floor, Operations can get very detailed data regarding the throughput of production lines, including any manual, test, and quality inspection operations. In addition, the MES can provide very clear and detailed data regarding the yield of processes, which can reveal constraints to the capacity of a production floor. The reporting of this data should be fed back into the enterprise resource planning (ERP) system for better planning and scheduling of orders (e.g., for procurement orders as a result of yield, vendor analysis, or production orders as a result of cycle time data). This information can also be used to develop better models of plant capacity that are used to improve sales and operations planning, provide greater visibility of investment needs and better capacity planning, and provide better visibility of the level of efficiency that can be improved. These improvements enable Operations managers to plan for improvement activities over a longer period (often as much as 6 months to a year into the future).

Operations' activities to improve the capability of a plant must be considered as an investment because expenses will be incurred in an effort to improve some part of product delivery. However, after the improvement has been implemented and a cause of variance has been eliminated, Operations must run with the new condition for a period before the cost of the initiative has been recovered. This is an element of continuous improvement (CI) initiatives that is often ignored. If the improvement is not sustained, the cost of the initiative will never be recovered and will result in a net loss for the plant. This is one of the main points of the Six Sigma methodology: ensure that the estimated return covers the expense of the implementation and that the actual outcome was achieved as expected.

Again, using an MES can help. First, by providing input on the actual activities that are affected by the original variance problem, Operations gains a greater understanding of the magnitude of the problem. Second, an MES functions as a process management tool to ensure that the process change is sustained, thereby helping ensure recovery of the CI initiative's cost. When the MES implementation is accompanied by implementing a Center of Excellence (CoE), the lessons learned from an initiative in one plant can be applied in all other plants. If the plants use similar processes and are manufacturing the same products, this transfer of knowledge is simple. However, there are additional opportunities for any process/product combination with similar operations. Although there will be additional costs to adapt the new solution for other products and processes, by making the initial investment, the plant will reduce the cost of implementing similar improvements at other plants.

Because the process change has already been stabilized at a previous plant, the return on the investment should be quicker. Although quality management and facilities management have very different goals, from an operations management perspective, the data analytics of each and the benefits of using the data are very similar.

13.1.2 Summary of Strategic Planning

It has been said that *if you fail to plan, then you are planning to fail.* This is particularly true in fast-paced international markets with data being shared across the globe in seconds. The issue is, what are you planning for? In today's world of global economics, the key to planning at the plant level is not only having access to relevant information about a wide range of a plant's performance criteria (throughput, yield, direct labor costs, etc.) but also ensuring that this information is available to all levels of the enterprise from senior management (for sales and operations planning) to lower-level management (e.g., key performance indicators that are used to monitor the actual needs related to all this planning).

A considerable amount of this information can be derived from (and is most accurately portrayed by) the data collected within an MES. The plant must also ensure that the information from the MES data is available throughout the enterprise (as many companies are picking up on the concept of "from the shop floor to the top floor"). When looking at the requirements to implement an MES, the steering committee must include an insight as to the type of plant performance information expected from the MES program and what other management systems will need access to this data. Keeping the entire data map of the enterprise in mind will help ensure that these expectations are met by the application and implementation teams. Many companies that have already implemented an MES did not have this scope of functionality in mind during the original request for proposal (RFP) process and spent a considerable amount of time and money developing their requirements based purely on manufacturing floor-level capability and IT requirements for server configuration and development support. Although these plant-level requirements are very important (and may be "make or break" deliverables), the ability of the MES implementation to support senior management planning (as covered in Chapter 11, "Developing an MES Strategy") should never be overlooked.

After an MES application has been implemented, action must be taken to ensure that the reporting from the MES is tied back into the strategic planning processes right up to senior management. Again, the information from an MES must be validated as part of the strategic planning process, and once validated, this information must be integrated into the senior management planning systems.

13.2 Gathering and Analyzing Requirements

The most important part of the implementation process is to properly gather, document, and analyze the system requirements. Without this step, anything that is developed runs the risk of not fulfilling the needs of the manufacturing activity. A common mistake is for Operations to select an MES application because someone has developed a relationship with one of the MES vendors and/or the implementation team, and a requirement document is created that basically states, "We want this system." The main issue with this method is that the "We want" aspect of the application is usually based on a limited number of pain points that the application may handle without a full scope of requirements needed to support the enterprise. Decisions made in this manner are usually shortsighted and can be problematic in the long run.

The process of developing a requirements package warrants a considerable amount of detailed analysis and understanding of the processes (indeed, "understanding the processes" has been a recurring theme) that the application must support and the data that must be managed through either data collection or reporting. As a result, the general outline of activity needed to create that understanding warrants a methodical approach to requirements gathering. As such, methodologies similar to that defined by The Open Group Architecture Framework (TOGAF) or other enterprise modeling methodologies are good models to follow. Using the full TOGAF methodology would likely be considered overkill for an MES implementation; however, an overview of the TOGAF process can be used as a guideline for the requirements definition process. The general outline of the TOGAF structure uses the following major points:

1. Create an understanding of the architecture (in this case, a MOM architecture) consisting of:[1]

1. Included in these architecture mappings are any regulatory and industry best practice requirements that must be supported.

- Business Process Architecture
  - This is a map of the manufacturing processes and the supporting business processes (in MOM terms).
  - From this, all the functional requirements for the manufacturing operator and the business support functions can be derived.
  - As noted in earlier chapters, using the full ISA-95 Activity Model (Part 3) as a template can help to speed up the process mapping and can help to ensure all processes are fully mapped.
- Data Model Architecture
  - This is a map of the data necessary to support the business processes.
  - This includes the data that represents the monitoring of the production floor (manufacturing processes, product traceability, quality management, equipment performance parameters, etc.).
  - This also includes the data needed from any applications that must be interfaced with. This will include determining which applications will "vault" the data objects.
  - Also, this will include database structures (current and needed data warehouses) and distribution.
  - From this, a business can derive the data the application will need and the interfaces that will be required.
  - Using ANSI/ISA-95.00.02-2018, *Enterprise-Control System Integration – Part 2: Objects and Attributes for Enterprise-Control System Integration,* can help to ensure the full reference of data.

2. Create a technical architecture of the MOM system that will include the following:[2]
   - All applications that need interfacing at the MOM level (defined by the data vaults)
   - All server requirements and operating systems (including cloud/hosting requirements)
   - All shop floor equipment that must be connected to the application and the type of interface requirements

---

2. Many companies have made it a hard requirement for either cloud or internal hosting for the implementation. As technology in this area develops rapidly, it is recommended that, at the RFP level, the selection group remains open to options.

After creating these two general mapping documents and the outline of a technical architecture, the actual requirements can be determined and documented. As stated previously, it would be unwise to expect the MES application to be installed without customization. As such, if there are any specific software development requirements (programming languages, development tools, etc.) as a result of other applications that have already been implemented, these should be included in the requirements document as well. If the implementation is part of a greenfield project or a brownfield update, the timelines and other project requirements should also be included.

The requirements document should be created without reference to how functionality and data structures are to be implemented, the purpose being that the implementation provider can present their ideas on how the application should work. This also functions as a test of the implementation provider's capability. To ensure a well-defined system, implementation providers must show that they understand not just the system but also the manufacturing environment the system is going to support. Failure to do so will likely result in solutions that do not fit the specified requirements (and can potentially provide misleading information) and are not designed with enough configuration ability to be applied to multiple plants.

13.3 Evaluating Solutions

After all the requirements have been documented, they must be organized and reviewed. Some requirements may come into conflict with others; accordingly, some requirements must be prioritized over others, and where conflicts are identified, they must be resolved or mitigated. The approval process must include all the stakeholders of the MES application and project. After the requirements document has been reviewed and approved by all stakeholders and the functional requirement specification (FRS) has been created, the next action is to send the FRS (as part of the full requirements document) to prospects as part of an RFP. The review and selection process for an MES is determined by a company's RFP process and is outside the scope of this book. However, as the scope of an MES is different from all other enterprise applications, if the company is not experienced in MES, it would be wise to enlist a consulting firm to help navigate the RFP process.

The approved FRS is then used as the basis for evaluating the various MES solutions available in the market. This can be a daunting task due to the multitude of solutions out there. Many solutions were developed to provide a certain niche MES function. These include systems that, for example, only support just-in-time or just-in-sequence order execution support or systems developed to support only a specific type of manufacturing, such as food and beverage or pharmaceuticals. These solutions may contain some of the specific functions detailed in government or industry regulations, enabling an easier deployment.

In choosing a solution, the following criteria should be considered:

- Can the system provide the required functions?
- How much of the system is a matter of configuration? Will some specific coding development be necessary to fulfill the requirements?
- What is the end cost of the system?
- How difficult will it be to reconfigure the system in the future?
- Depending on the complexity of the system, what are the ongoing support costs?

An important consideration is the objectivity of the vendor developing or providing the solution. The vendor must be attentive to the details in the FRS and ensure that the solution provided will meet those requirements. Too often a vendor provides their branded solution rather than tailoring it to the requirements laid out in the FRS. A technologically agnostic MES solution provider that does not promote any specific solution over another can help provide such an objective solution.

After all these considerations, it may make sense to either upgrade the current system (if there is one) or go with a unique homegrown system. Ultimately, the choice will depend on the cost of ownership over the entire lifecycle of the system, which includes ongoing support costs and estimated costs for future configuration changes.

13.4 Understanding Data within Operations (Reprise)

Operations management has been an important discipline for decades, and in the manufacturing industry, MOM, in particular, has become an important discipline worldwide as manufacturing supply chains have created considerable international distribution routes. The function of MOM is to examine plant capability by analyzing data (ship quantities, yield rates, cycle times, amount of variance from the norm, etc.) and to examine the relationship of these data to the production process (production operator activity, delivery of material, equipment performance in relation to product quality, etc.). In studying these relationships with exact definitions of magnitude (e.g., how much, how long, and how fast), operations managers work to improve the effectiveness of their manufacturing processes to achieve lower costs of goods sold and still maintain acceptable profit levels. Operations managers measure their own management capability (i.e., just how good of a manager are they?) by these data. An MES application is the one tool that has been explicitly designed to help MOM accurately monitor and actively manage the plant's capability.

ERP software applications were used as early as the 1960s (when you include the use of ERP's predecessor, material requirements planning—MRP). They were designed to plan and monitor the conversion of anticipated/actual customer orders to delivery and to simultaneously plan and monitor the material and equipment requirements needed to fulfill customer orders and convert this information to monetary values. The issue with ERP systems is that the planning and monitoring involved are also accomplished at aggregate and summarized levels. With the requirement of computing power necessary to perform this function and the inherent delay in the data available for ERP (as shown in Chapter 2), the aggregate and summarized levels were the best level of data calculations that were capable at the ERP scope of function.

Programmable logic controller (PLC) systems, alternatively, were designed to automate the manufacturing process by having work performed in a repeatable and cost-effective manner. The aspect of monitoring equipment performance arose from a need to capture streams of sensor data for troubleshooting and focuses on specific characteristics of that equipment, with the expectation that capturing variances in that data would point to areas of concern in equipment performance. As variances in sensor data at the level of a tenth of a second could be significant, historian databases were designed to capture highly specific data at rapid speeds, but they only related the data to the specific time of occurrence. As a result, historian databases are unable to link this data with other events that may have a relationship to the equipment performance. In addition, PLC and supervisory control and data acquisition (SCADA) systems were designed either to view specific equipment performance or to monitor the overall process performance, but only at the equipment level. The technology for this is not conducive to tracking material and production operator interactions outside the scope of the equipment level.

MES applications were designed for the specific purpose of collecting data from equipment, operators, and other systems at speeds that are fast enough to be meaningful and to store that data in a manner that maintains the relationship with other events occurring on the manufacturing floor. The function of MES applications has been around since the late 1990s, and since that time, MES have been studied, analyzed, and improved to support this key function. MES have been praised for their contribution to MOM and scoffed at by those who do not understand the MES concept and the value that is provided. Presently, some see the Industrial Internet of Things (IIoT) as the next stone to be cast at MES that will bring them all crashing down.

There is, however, one irrefutable fact about MES: with the right application and the right implementation method, an MES will provide visibility to where manufacturing is having problems and will highlight the magnitude of those problems fast enough to enable production operators to make a difference before a problem gets out of hand. An MES will also provide that visibility in a much more succinct manner than just about any other manufacturing operations tool. These MES visibility factors are important in equipment uptime, product quality, and production metrics.

There have been attempts from both PLC and SCADA providers, and from some ERP providers, to reach into ISA-95 control Level 3 that is currently supported by the MES. In each of these cases, the technology of the designs failed to provide the capability needed at this level, or the vendors designed separate applications with MES capability and tried to brand them as control-level applications, not MES. However, because customers compared these applications to MES applications, they were competing as an MES application.

13.5 Conclusion

With the new technology of cloud, edge, and fog applications and the development of IIoT technology (e.g., block chain and possibly autonomous MES), the face of MES will certainly change. However, the functionality that an MES provides will always be a requirement for MOM as long as there is a possibility of a less-than-perfect process, using less-than-perfect material or equipment, and creating a less-than-perfect product.

We created this book with the intent of getting past the marketing hype that vendors spin and the scuttlebutt that naysayers present. Our main goal was to relate the concept of MES to a real-world, operational point of view. We would like to hear your feedback and welcome the opportunity to discuss differing opinions.

Appendix: Introduction to Autonomous MES Computing

Introduction

This appendix introduces the concept of an autonomous MES architecture. An autonomous MES is not a proven concept, and the white paper will be updated as the concept becomes better defined through discussion streams via social media. The goal of this document is to aid in the development of the autonomous MES concept and to provide a sounding board for the development of an open-source standard communication protocol that can be used as the basis of a "services"-based MES.

Problem Statement

Many manufacturing companies use several IT systems that collect, aggregate, and process data and transactions to keep track of sales, inventory, and shipments, as well as a whole host of other data objects, and then generate reports for information that enables companies to understand their current performance. One of these systems is a manufacturing execution system, an MES.

The MES is designed to lead production operators through the manufacturing process as production units or work-in-process (WIP) move through that same process as defined by a company's manufacturing engineering. As part of the function of an MES application, the system records all activity that was performed on each unit of WIP and maintains the complete as-built history or device history report (DHR) of each unit of WIP. All of this data is maintained in central databases in order to provide the reports and information that people need from the system and to provide guidance to production staff related to the up-to-the-second status of the production floor. In most cases, MES databases must accurately record and provide feedback on transactions to the latest second (sometimes even into the millisecond range).

An MES is also "aware" of what the process should be and what components should be used to build the product. In many implementations of an MES, the system is required to verify many (if not all) of the steps needed to manufacture a product before each step is actually performed. The system also must take into account what equipment is being used at that moment in time, the current state of process conformance for any particular production unit, and the "up-to-the-second" current state of material availability. The result of this verification by an MES is that every transaction (i.e., every operation performed, component assembled, and test performed) must be verified as the correct action at that moment in time before that transaction is to be performed.

The outcome of all of this verification is that all data and transactions must be transmitted, verified, and recorded in the central database in real time.

With a small production demand of a 20-step process, 100 components to assemble at a line rate of 1000 units per hour comes down to about 40 transactions per second for that line. Within each production transaction, the MES will perform three to four verifications, giving 120 to 160 application transactions per second per line. This does not include the movement of data that is generated by any of these transactions. In an environment such as this, a conventional MES has no problem maintaining that rate of production. However, when production areas have 25 or more lines with greater than 50 steps and hundreds of components, the demand for communication can grow very rapidly.

If you were to take a look at the data traffic from a networking perspective, as the segments of data (called *packets*) move through the network, the closer the packets get to the main MES server (or cluster of servers), the more each transaction must compete with other transactions to be completed in the required second to subsecond time frame (packets communication becomes denser the closer you get to the target server).

In the manufacturing sector, the Industrial Internet of Things (IIoT) is on the rise as more equipment used in production is machine-to-machine (M2M) enabled (i.e., the equipment has a considerable capability to perform computational transactions and provide greater granularity of process characteristic-data to the MES environment). As this trend in data and information availability grows, it places more and more demand on the MES infrastructure to receive, process, and act/report on the current status of production. It is conceivable that this demand can and probably will create bottlenecks (as shown in Figure A-1) in the communication between the MES server(s), the workstations, and the operators performing the work. This bottleneck would then create delays in the processing of MES transactions in an environment where true real-time communication has been proven critical to providing information to the production operator at a rate that does not slow the operator down. As MES are implemented, many consulting companies are already finding it difficult to provide a system response that does not impede the speed of production.

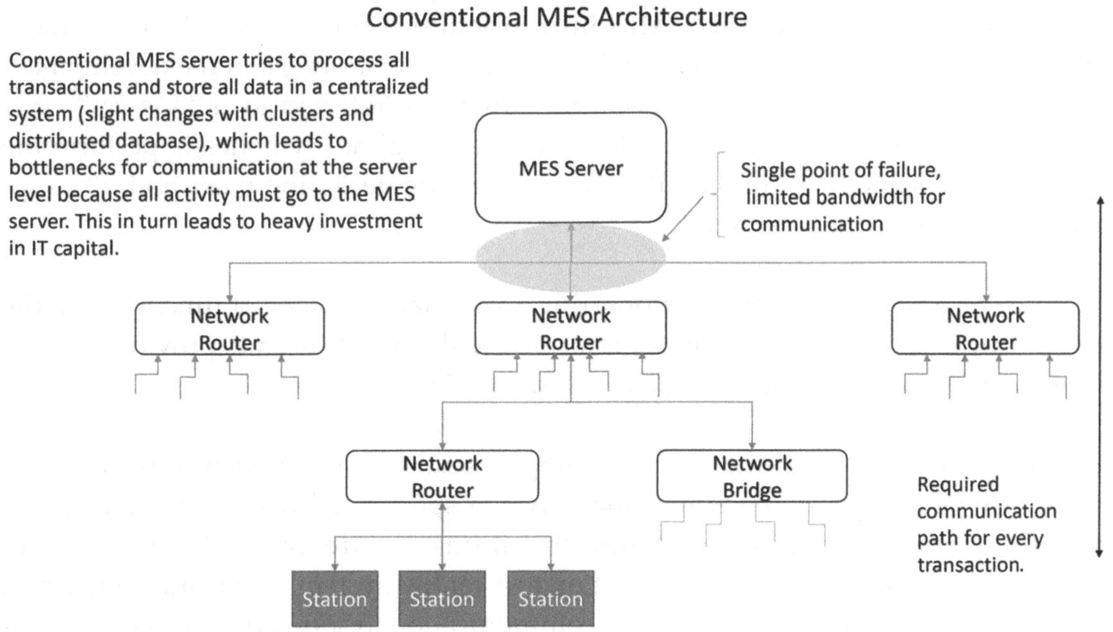

Figure A-1. Conventional MES architecture (constraint).

In an attempt to resolve this problem, many application developers are implementing applications using cloud computing systems. This method of implementation can provide some relief but does not actually resolve the bandwidth constraint issue. In many implementations, it is the network infrastructure that causes delays and not the MES application itself. The MES can be the fastest in the world and still be limited by the bandwidth of the network. In order to resolve the communication constraint, a new infrastructure for the MES must be developed.

White Paper: Autonomous MES

Introduction

In recent years, the computing power of devices has grown, and that computing capability has been made available in smaller and smaller packaging to the extent that the common functionality of a smartphone now has greater computing capability than a standalone computer from a couple of decades ago. In addition, more diverse communication paths have now been made available within the packaging of a smartphone device. These communication paths include Wi-Fi, Bluetooth, and cellular and near field communication (NFC), all standardized within a single package. With some redesign, additional communication paths can be developed, including designs in spread-spectrum FM communication.

By using devices similar to smartphones and the capabilities being developed in IIoT devices, the introduction of a new MES infrastructure is now possible. By using these computing devices (referred to within this document as *"nodes"*) and the development of small programs to perform very specific functions (referred to as *"services"*), it is now possible to create a new MES infrastructure using mostly nodes to perform the specific tasks needed within the MES environment by designing these nodes to support peer-to-peer communication.

Because these nodes will only be supporting a very narrow scope of functionality, the speed with which these tasks can be performed will make it easier to provide the real-time speed of communication needed within an MES.

As shown in Figure A-2, the proposed implementation of this new infrastructure would use peer-to-peer communication between specialized nodes for the majority of transactional communication. Within this infrastructure, the nodes engage with each other directly without communication to a central system. The result would be a significant distribution of the necessary communication to subsections of a company's

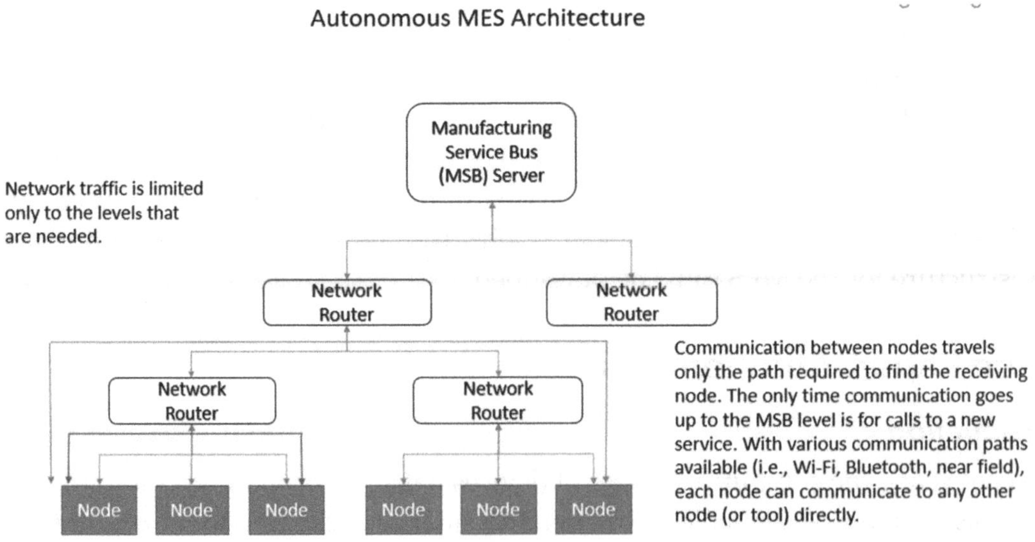

Figure A-2. Network communication—autonomous MES.

network, thereby reducing the bandwidth demand within the network that is currently seen in conventional MES infrastructures.

Components of the New MES Infrastructure

Manufacturing Service Bus

The manufacturing service bus (MSB) is the only part of this concept that maintains the centralized computing structure. The purpose of the MSB is to provide a repository for services to the nodes and to maintain and process the MES activities that are not of a transactional nature. These MES activities would include functionalities such as reporting, data aggregation, and event management. Further description of these functions will follow later in this document.

Computing Nodes

Computing nodes are small computing devices with the capability to communicate over multiple communication paths, such as Wi-Fi, cellular networks, Bluetooth, and NFC infrastructure. These nodes could also be designed to handle Ethernet communication, if required. Further options for communication may include other spread-spectrum communication protocols yet to be developed.

The purpose of a node is to act as an autonomous computing device that is mounted as an integral part of equipment on the production floor such as a workstation (the location on the production floor where work is actually performed), product carrier (equipment or containers that are used to move units that are in production—WIP—through the production process), or line-side stocking (LSS) unit (locations used to maintain component inventory needed for the production of units). Within the concept of the autonomous MES, it is expected that there could be hundreds of nodes within a production environment, all performing specific tasks within the MES as a whole.

As part of configuring computing nodes, each node will be required to maintain its own IP address, its purpose within the MES architecture (i.e., what MES function it will be performing), and where it fits within a master node configuration.

Computing Services

Computing services are small, very specific applications that have been designed to operate at the node level in an operating environment designed specifically for peer-to-peer communication. These services must be designed to function within an operating system similar to that of smartphones and have the look and feel of an app running on a smartphone. Each node would have one or more services to provide the peer interaction necessary within any specific MES function. Additional functionality must be embedded into each node to provide the capability for each node to communicate to other nodes over the various communication channels. These services would also include application support to communicate with other equipment within the defined communication protocols. This may include communication with test and sensory equipment, automated actuators, or other tools needed to perform specific tasks.

Each service would maintain the data locally (on the computing node) that is specific to that service for the duration of the time that a production unit is considered WIP. Once that unit is complete (either to finished goods or scrapped), all of the data for that unit would be aggregated and transferred to the central data warehouse relieving the services from the responsibility of holding that data.

Some of the tasks needed in normal MES functionality would include the reporting, data aggregation, and event management, introduced earlier. The services to provide the necessary support for these functions would operate within the MSB infrastructure making them available for use within a cloud environment.

Sample Services

The remainder of this white paper is dedicated to providing a description of specific services that would be required and to describing sample interactions of MES activity to provide a sense of how the system would operate. The examples described are selectively stated for

a discrete manufacturing environment. The discrete manufacturing environment is used only because of familiarity to this manufacturing model by the author and is not intended to limit the scope of functionality of this new MES infrastructure. Any support of manufacturing environments other than discrete manufacturing must be developed as the demand arises.

Carrier Service

The carrier service would be one of the primary services needed within this new MES. It would reside with the product carrier node, and the local data storage would be the single reference point for all device history data for the unit associated to the carrier at the time of production (i.e., as a product moves through the production process, each unit would be assigned to a carrier and therefore the carrier node). Some of the data objects needed as an integral part of this service are:

- **Unit identifier** – The unit serial number (or batch number) used to uniquely identify each particular unit of WIP on the production floor.

- **Part number** – The system part number that identifies the product being built within the production order.
- **Manufacturing bill of material (mBOM)** – Contains all the raw material (components) needed to manufacture a single unit of the product being manufactured, herein referred to as *WIP*. It would also contain a reference to the operation within the build process where each component would be assembled, any alternate components for that product, as well as a reference to the type of data that is to be collected by the operation at the time of assembly.
- **Bill of process** – The complete list of all operations needed to complete the build of a product (referenced herein as the *build process*). In manufacturing environments with more than one production station that can perform a specific operation, the build process may need to include a reference to the workstations that can perform a particular operation.

Assembly Service

The assembly service is used as the primary function for assembling components to the unit of WIP while that unit is being worked on at a specific workstation. The workstation would be configured to have a node specifically associated to the physical workstation, and the node would be defined to use the assembly service. The assembly service would be defined to know the operations that can be performed at that workstation.

The data objects needed as an integral part of this service are:

- **Resource** – The major capital equipment identifier that defines where a particular operation is performed. In some production environments, the operation and the resource may in fact be the same piece of equipment (this is common, for example, in the automotive industry).
- **Operation** – The particular part of the manufacturing process that identifies what will be performed. Within some manufacturing environments, a single station may perform multiple operations. The operation should relate directly to a step in the manufacturing process for a specific unit of WIP.
- **Operator** – For an operation that is not completely automatic, there will be an operator (or multiple operators) who is expected to perform the operation. The operator should be linked to a specific operation and resource combination.
- **Work center** – Depending on the definition of the production floor, a workstation may be associated to one work center with the possibility of having layers

of additional work centers in a hierarchical structure. A resource would be associated to a work center and the work center associated to a higher-level work center.

- **Tools** – Within the structure of a station, a particular resource may have a set of tools associated with it. Each tool defined within the system should be associated to only one resource.

Display Service

As the assembly node (or any other functional node) is defined for both automatic and/or manual workstations, it is possible to have nodes defined that do not use a display unit. As a result, it may be more effective to define the display node as a separate service and have the display node interact with the assembly (or other functional) node. The interaction between the display node and the functional node would depend on the type of information required to be displayed at a particular point in the process. In this regard, there will be a need for function-specific display nodes or the basic display service to be able to reconfigure itself depending on the functional node with which it is interacting.

Having a separate display service enables the functional node to interact with the display device being used. The functional node would provide consistent data to the display node according to a standardized peer-to-peer protocol, and it would be the responsibility of the display node to handle the required data. It would also allow the display node to pull other information as required from other sources (also on a peer-to-peer protocol or via the MSB) without placing computing demand on the functional node.

LSS Services

In this context, LSS services provide the function of handling line-side inventory as part of a workstation configuration. Using human-machine interface capabilities, such as "pick-to-light" and the interaction with other services, the LSS service would track and record the current inventory levels at each LSS location and maintain local management of the LSS inventory as it is consumed. Additional functionality of the service as part of the management of local inventory levels would be to issue signals to the MSB and to the display service in the event of low inventory levels at that location (or a stockout for a particular inventory bin at that LSS location). At a local level, the data objects required include:

- **Workstation or LSS location** – The data object that defines where on the production floor inventory for the node is to be found or delivered.

- **Inventory bin identifier** – A unique identifier is required for a specific holding bin or feeder shelf for a specific component part number that is being maintained at the LSS location.
- **Component part number** – As with standard inventory management practice, material would be expected to have a unique identifier for each part within the materials list. This material identifier would be matched to the same identifier used by ERP systems as part of an inventory management strategy. There may be multiple part numbers in inventory at that LSS location depending on the strategy or production process.
- **Traceability identifiers** – As part of the traceability requirement, most manufacturing companies require either a batch ID or a serial number to be collected at the time of assembly of components to the manufactured product, thereby recording the genealogy of the as-built product. When material is moved to the LSS location, the inventory operator would be required to identify the batch or serial numbers that have been delivered to that location. In some instances, when the inventory operator delivers a new bin to the LSS, the LSS service would communicate with a radio-frequency identification (RFID) tag on the bin, which contains both the inventory bin and the traceability identifiers, permitting each to automatically communicate this data to the LSS service. The LSS service would be responsible for holding the unit identifiers that have been stored at that LSS location.
- **Min/max levels** – As part of the strategy for maintaining LSS inventory at optimal levels needed to meet a production schedule, the min level of inventory for that LSS location would be used to trigger an inventory replenishment event, and the max level would be used to identify how much should be transferred to that location. In many cases, the max level can be predefined at the ERP level as the standard fill quantity for the bin when a transfer is needed to the LSS location.
- **Current inventory level** – As material is consumed from the LSS location during assembly, the LSS service would maintain the current inventory level.

Sample Services Interaction

The following are examples of the communication sequence between services, both as a basis for dialog on what the design of such a system would require and as a foundation for the design of a standard protocol to allow these services to interact. The design of a communications protocol is beyond the scope of this white paper, but the author hopes that other organizations will be inspired to develop such a protocol.

Because many of the interactions between nodes take place on a peer-to-peer basis, the initiation of internode communication must include the IP address of the nodes as they initiate communication with each other. As communication between nodes over Wi-Fi must have the IP addresses to start communication, it is recognized that as part of the protocol for communication, the initial communication will probably need to be engaged through a different path (e.g., Bluetooth or NFC). This initial communication will require a handshake between nodes, and each node then must configure its communication requests to communicate with the IP of the other node.

Once the handshake process has been completed, communication between the nodes via Wi-Fi or Ethernet for the much larger amounts of data needed will be exchanged between nodes to complete the transaction.

MES Assembly Interaction

In this interaction, the services that will be participating for this example (referred to as *actors*) will be the carrier node and the workstation node. Each would have been preconfigured to operate with the required services when the nodes were originally placed into service. The workstation node would also be configured to communicate via Bluetooth or an NFC sensor (part of the smart sensors talked about earlier in this document). As an option, the workstation node would also be configured to communicate with an LSS node located at the workstation, and the workstation would have a display node also configured for the operator.

When the NFC sensor detects the carrier node, the sensor triggers the workstation node to engage the carrier node; through the NFC sensor, the carrier node and the workstation node perform a handshake and initiate communication via Wi-Fi.

Once Wi-Fi communication is verified, the carrier node recognizes the operation that the workstation node is to perform, and the carrier node provides the workstation node with the list of materials required to assemble (including part count and possibly data collection requirements).

At this point, several interactions could take place:

1. The workstation node communicates with the LSS node to ensure that the required material is available.
2. Then the workstation node communicates with the display node to inform the operator of the list of materials required to assemble.

3. The LSS node shows the operator which material to pick first (using pick-to-light) and indicates the pick sequence for the remaining material.
4. As the operator picks each piece of material and assembles it, the workstation node communicates to the carrier node and to the display nodes to update the list of assembled items.

Once the assembly process for that workstation is complete, the workstation node and the carrier node perform a final handshake to ensure all data has been updated, and the two nodes end their peer-to-peer communication. The carrier node will then be moved to the next workstation to perform the next step in the process.

If the assembly process included the use of tools, each of the tools would be preconfigured to the workstation and could communicate with that workstation over Bluetooth or peer-to-peer via Wi-Fi for intelligent tools.

Inventory Replenishment Interaction

In this interaction, the actors are the LSS node (which maintains the inventory of material used in the assembly process), the workstation node (where the LSS was configured), the MSB, and the ERP system managing the inventory centrally.

In this example, the LSS node is capable of communicating directly with the MSB and is already configured with the IP address of the server used for the MSB.

During the assembly process, it is often normal for the LSS station to have multiple bins of the same material available at the location in a first-in-first-out feeder shelf. As the assembly process uses the material in a single bin, it is possible for the LSS node to recognize when a bin has been removed by using bins with RFIDs. This event can be used to trigger a replenishment of the single bin. In this case, the LSS node would communicate the bin ID, the material, and the quantity that was originally in the bin (which now is assumed to be depleted) to the MSB, and the MSB would then send a replenishment signal to the ERP system to have the material bin replaced.

When the inventory handler comes to the LSS to replenish and update the inventory levels, the handler will use a communication method such as NFC to engage the LSS node to their display node (once again using a handshake), and the handler will place a new bin in the required bin location (which would automatically communicate with the LSS node) and acknowledge the completion of the material transfer. The LSS node would then update the status of local material to the MSB, causing the MSB to cancel the original "inventory replenishment" event that was triggered.

During the assembly process example above, if the on-hand quantity of material within a specific inventory bin goes below the minimum quantity configured within the LSS service, and as the LSS would probably have an inventory replenishment request already open, the LSS node would then initiate a "low inventory" event with the MSB. The low inventory may be due to multiple bins having been depleted before replenishment or the availability of only a single bin for the material. As part of that event, the LSS also signals the workstation node, and the workstation node communicates with the display node to inform the operator of the event.

In both cases, the MSB would perform a look-up within the event library to determine the actions needed. In each case, the MSB may perform a handshake with the LSS node to determine the specific material, the location of the node on the production floor, and the bin number that must be replaced. The MSB would then communicate with the ERP system to request the material replenishment and reconcile any discrepancies (as a result of scrapped material that was not assembled).

Functional Testing Interaction

In this interaction, the actors are the carrier node, the workstation node, and the test system. Similar to the assembly interaction, the process starts by the carrier node signaling the workstation node via NFC to initiate the handshake for Wi-Fi communication to begin. In this case, the test system would probably be a standalone computer system or a test server. As a result, the test system must have a client application that can simulate the exchange of nodes.

Once the test system has confirmed communication with the carrier node (note that some test systems require the tester to download the test sequence), the tester node initiates testing by providing instructions to the test operator to prepare the unit by connecting the unit to the test jig and signaling that the test is ready.

Note: Depending on the amount of computing resources needed, the test system may, in fact, still be configured as a node operating with a customized test service. Also note that the strategy for maintaining the actual test data may change so that the test system can upload the results of the test to the MSB if there is a significant amount of data to be stored.

If the test is successful, the operator will be notified to disconnect the unit, the workstation node (or test system) will update the product carrier with the results of the test, the final handshake to clear the test system will be performed, and the carrier will be moved to the next operation.

If the test fails, there are several strategies that can be utilized depending on the priority of the resources (e.g., is the tester a constraint, or is the product movement a constraint?).

Depending on the situation, the test node may be required to notify the MSB while the product carrier is released and starts moving (the movement of the product could be altered after release from the test system). Alternatively, the carrier node could be released from the test system, making the system available for the next unit, and the carrier node could then contact the MSB to handle the test failure event. The data regarding the test results would still be either transferred to the carrier node or uploaded to a central test repository until the carrier reaches the workstation where the nonconformance of the unit can be processed.

WIP Reporting Interaction

In this interaction, the major actors are the MSB as the primary reporting system and the product carriers that are currently active in production. The purpose of the WIP report is primarily to provide Operations with the current status of WIP within production, details such as the part numbers of all products in production, the breakdown of the status (active, scrapped, in nonconformance, on hold, etc.), and the breakdown of the age of WIP in production (this information may be provided as an aged WIP report). Depending on the implementation, other data may also be provided. This type of information is frequently included as part of a periodic reporting structure for production billboards to maintain a constant update of that information or as part of a takt board. In any of the above, the data required would be very similar and would be retrieved in a similar manner.

In this interaction, the report is initiated as a standalone request by an operations user. From a personal computer, the user opens an initiating page to access the MSB to reference the service list. From the service list (configured to that user's permission level), the user selects the WIP reporting service, causing the service to load into the standalone computer. This can be different from a conventional MES because even from the personal computer, the WIP reporting service may be a standalone application instead of the client/server screen used in most systems.

The user accessing the MSB selects the desired reporting service, the service downloads to the computer (if the service has been previously downloaded, this would not be required), and, once loaded, the MSB communicates directly with the service to perform limited configuration of the report service for that particular user/service.

The user then initiates the service on the computer, selects specific parameters for the report (typical capabilities may include filtering for a particular station, product, and/or time frame) and then requests the report. As the product carriers in production are the primary source of data for WIP, the report service must access that data via a reference to the product carriers. At this point, there are two options for activity depending on the manner in which the service is implemented:

1. The carrier list is transferred to the service.
   a. The service references the IP addresses of the carriers or a list of carriers that are currently active in production.
   b. The service then sends a data request to the carriers to receive an update on the latest status.
   c. If a special set of data is required for the report from a special source, the service requests that data as well.
   d. Each carrier responds to the request with a standardized response.
   e. The service at the computer level aggregates the data returned specifically for the report requested and presents that report to the user.
2. The service requests the report from the MSB.
   a. The service requests a WIP report from the MSB.
   b. The MSB determines the services needed to get the data for a WIP report.
   c. Using the IP list for carrier services, the MSB sends the request for data to the carrier services currently active in production.
   d. The carrier services respond with the required data to the MSB.
   e. The MSB aggregates the data from the carriers and forwards the aggregated data to the requesting service, which displays the aggregated data in the form of a WIP report.

In either case, there are pros and cons to the method of implementation; the determining factor is whether to place the load of aggregating data onto the service or onto the MSB.

Reporting Summary

With any other report, the interaction between services would be similar with either the MSB performing any required data aggregation or particular services being primarily responsible for maintaining the data needed to satisfy the reports. In many cases, the data required for one report is very similar to the data for others. What changes are the method of aggregation and the manner of display. If the MSB has already requested data from specific services, it may be possible to re-aggregate the data without a request to the production services. For this method to work, there must be an exchange from the services to the MSB when there has been a change in the status of the production service. Without this exchange, the MSB would be required to prompt the production services for every report request.

Services Interaction Summary

As noted in the MES functional interactions, for many of the MES functions there is a significant amount of communication that will happen in a peer-to-peer protocol. This change in the communication strategy provides a large redistribution of the MES's communication over the entire network and removes the constraint of a conventional MES that uses what is known as a *client/server communication protocol* back to a centralized server and database farm. As the industry adopts this kind of communication protocol, there are possibilities for savings as central servers do not need to be as powerful and loads on the network are more evenly distributed, causing less demand on the database, server farms, and network infrastructure.

Once the standardized protocol is available, it will be easier for companies to engage in the development of other services. Having new services available from central cloud distribution environments makes it possible for manufacturing companies to "subscribe" to the particular functionalities they need. In addition, this new infrastructure makes upgrading (a current serious pain point for an MES) continuously available because only the specific services that are due must be upgraded and, as long as the standard requires backward compatibility to any supported level of the standard, the MSB can be upgraded almost as easily.

The primary requirement of such an infrastructure would be the development of a standard protocol that would be used industrywide.

Industry Adoption

The proposed infrastructure would be developed to operate on an industry-standardized communication protocol and industry-normalized data objects (for example, using ISA-95 and Manufacturing Enterprise Solutions Association's Business to Manufacturing Markup Language as models).

Using this protocol, manufacturers of product carrier systems, workstation equipment, and so on could introduce equipment that already has nodes embedded within their equipment and provide services to support that equipment as an integral part of their product line.

Current system providers would develop their systems to provide the infrastructure communication (i.e., MSB) to permit the nodes to communicate with the enterprise system as a whole and to maintain the reporting, event management, and services repository, and would develop many of the common services that would be needed within the MES infrastructure.

Additionally, by having the infrastructure designed to support a standardized communication protocol and based on industry-normalized data objects, a market would then become available for developers to create services to access, interact, and manage data similar to the apps on smartphones.

This infrastructure can be used as the primary platform for developing the future of intelligent manufacturing systems, currently being discussed as part of Germany's industry 4.0 and America's Smart Manufacturing initiatives.

Summary

In recent years, many companies have been looking to invest in the market being created by the IIoT. The problem within the manufacturing industry is that there seems to be very little effort to design a standard protocol for communication. This leaves system and product providers with the dilemma of supporting multiple systems to make their products compatible with industry leaders. With the development of an infrastructure similar to what has been defined within this paper, the industry as a whole could define, develop, and implement a single broad strategy that would be able to grow and scale as required, regardless of the demand that is being placed on MES from M2M or IIoT initiatives.

Bibliography

ANSI/ISA-95.00.01-2010 (IEC 62264-1 Mod). *Enterprise-Control System Integration – Part 1: Models and Terminology*. Durham, NC: ISA (International Society of Automation).

ANSI/ISA-95.00.02-2018. *Enterprise-Control System Integration – Part 2: Objects and Attributes for Enterprise-Control System Integration*. Durham, NC: ISA (International Society of Automation).

ANSI/ISA-95.00.03-2013. *Enterprise-Control System Integration – Part 3: Activity Models of Manufacturing Operations Management*. Durham, NC: ISA (International Society of Automation).

ANSI/ISA-95.00.04-2018. *Enterprise-Control System Integration – Part 4: Objects and Attributes for Manufacturing Operations Management Integration*. Durham, NC: ISA (International Society of Automation).

ANSI/ISA-95.00.05-2018. *Enterprise-Control System Integration – Part 5: Business-to-Manufacturing Transactions*. Durham, NC: ISA (International Society of Automation).

ANSI/ISA-95.00.06-2014. *Enterprise-Control System Integration – Part 6: Messaging Service Model*. Durham, NC: ISA (International Society of Automation).

Vokey, G. (2018). "White Paper: Introduction to Autonomous MES Computing." White paper, Rockwood, ON: Vokey Consulting, Inc.

Glossary

blue printing. The part of an implementation project that analyzes the current state of a manufacturing plant's operation and determines the vision of a future state. It compares the current state to the vision of the future state and creates a roadmap of changes to achieve the future state.

BoL: bill of lading. A shipping document used to describe the contents of a truck or other shipping container and the formal document of receipt for the carrier.

BOM: bill of material. A list of materials and the quantity of each that are used to make one unit of a product.

BOP: bill of process. A list of operations and/or tasks that must be performed to make a production unit.

computing node (node). A small hardware-computing device that is implemented as an integral part of the manufacturing equipment in production and used for specific application support.

computing service (service). A small software application designed to operate within a node or on some other computing device that operates as an integral part of the autonomous operating infrastructure.

CRM: customer relationship management. The aggregate of processes, systems, and data used to manage the ongoing relationship between a company and its customers.

DFM: design for manufacturing. A design methodology that focuses on designing a product in a manner that optimizes the capability to manufacture the product.

ERP: enterprise resource planning. The aggregate of processes, systems, and data used to manage the ongoing capabilities of a company as a whole (or enterprise).

GMP (cGMP): (current) good manufacturing practice. A prescribed set of manufacturing methods and practices used by several industries (medical devices, food and beverage, cosmetics, pharmaceuticals, and health supplements) to ensure the quality standards of the product and provide regulatory traceability.

IIoT: Industrial Internet of Things. The interconnection of sensors, instruments, and other devices networked together with industrial applications running on various computing devices to support manufacturing, energy management, and other industrial purposes.

IoT: Internet of Things. An extension of Internet connectivity into physical devices and everyday objects (cyber-physical devices). These devices are embedded with electronics, Internet connectivity, and other forms of hardware (such as sensors); they can communicate and interact with other devices over the Internet and can possibly be remotely monitored and/or controlled.

ISA-95: International Society of Automation ISA-95 Series of Standards. These standards contain an IT model linking the processes of the shop floor to the top floor of manufacturing.

JIT: just-in-time. A process in which manufacturing scheduling is tightly managed to minimize any gaps in time from the launch of production and delivery of any unit to the expected demand of the end product.

LSS: line-side stocking. An inventory stocking methodology that places small quantities of material stock at a workstation-level inventory location. It is used to ensure easy access to material that will be needed within a short time window.

M2M: machine-to-machine. Early concepts of IIoT used to reference communication events directly between two or more machines without a central database system.

mBOM: manufacturing bill of material. A list of the raw material (components) required to manufacture a single unit of the product including data that is important to the manufacturing process such as operation of consumption, process, or genealogy data to be collected.

MES: manufacturing execution system. A system used to monitor and control the activities and resources of the manufacturing floor within a company (or enterprise).

MESA: Manufacturing Enterprise Solutions Association. An international not-for-profit community dedicated to improving business and manufacturing practices through optimized application and implementation of information technology (IT) and best management practices.

MOM: manufacturing operations management. The aggregate of processes, systems, and data used to manage the ongoing activities on the manufacturing floor.

MSB: manufacturing service bus. A services support infrastructure designed to manage the availability of services to computing nodes, manage the processing of events initiated by services, and possibly perform aggregation of data sourced from services and used in reporting.

NPI: new product introduction. The aggregate of processes, systems, and data used to manage the ongoing activities of bringing a product from concept to market.

OEE: overall equipment effectiveness. A manufacturing management key performance indicator (KPI) used to measure the current effectiveness of a manufacturing resource in delivering *good product* in comparison to the resource's full potential.

OEM: original equipment manufacturer. A manufacturing organization that makes a consumer end product from parts and subassemblies ordered from other manufacturing organizations.

OT: operational technology. From a manufacturing perspective, it is the hardware and software that directly acts on a production unit as part of the manufacturing transformation process.

plant. Manufacturing facility. A reference to a building (or multiple buildings) in which manufacturing takes place.

PLM: product lifecycle management. The aggregate of processes, systems, and data that are used to manage the ongoing lifecycle of a product from concept and release through to product obsolescence and discontinuation.

product carrier. Equipment, such as conveyors and lifters, designed to hold and carry units of work in process (WIP) as they move through the production process. Product carriers can be used in an automated or manual production process.

readiness. The status of equipment, material, or process that references its availability, operability, and suitability for use. This can include scheduled availability, having an operator available to use it, and the "fit for use" status for the task, including calibration/maintenance and the capability to perform as required.

RFID: radio-frequency identification. The technology of encoding digital data (barcodes) onto tags or smart labels for the purpose of tracking the items to which the tags/labels are attached. An electronic reader is used to access the data.

SCM: supply chain management. The aggregate of processes, systems, and data that are used to manage the ongoing activities within an enterprise for the receipt of raw materials from vendors to the delivery of end product to customers. (Note: In some situations, this may include the extended supply chain over multiple companies.)

TOC: Theory of Constraints. A method of system and process analysis that states that all processes will have a single constraining function and prescribes methods to manage and improve a process based on the defined constraint.

WIP: work-in-process. One or more units of a product in the process of being assembled or fabricated.

Index

Printed in the USA/Agawam, MA
September 17, 2026

910661.227